the
SOLAR
ELECTRIC
INDEPENDENT
HOME book

SECOND PRINTING
REVISED EDITION
COPYRIGHT 1993

Written and Published by
Fowler Solar Electric, Inc.
226 Huntington Road
Worthington, MA 01098

the SOLAR ELECTRIC INDEPENDENT HOME book

Published by: Fowler Solar Electric Inc.
226 Huntington Road
Worthington, MA 01098

413-238-5974

ISBN 1-879523-01-9

Library of Congress Catalog Card Number: 91-71903

TABLE OF CONTENTS

Introduction

This book is a product of our years of living with solar electric systems, and designing and installing solar electric systems. The owners of the remote PV homes of today are really the pioneers for this new and promising technology. We hope you will view your PV home as a model for PV homes to follow and therefore spend a little extra time and a little extra money to do the best job possible. We hope you will bring your system up to the standards of the *National Electrical Code®* (or *NEC®*)*, have it inspected by the local electrical inspector, show it to everyone you meet, and spread the word that there is a new, better, and independent way for us to produce our home electricity.

We, the staff here at Fowler Solar Electric Inc. have strong science, electrical engineering, and environmental studies degrees. We have taught these skills at many levels. Furthermore we have practical experience in electrical wiring, electronics, homebuilding, and just plain being independent homesteaders. We are strong advocates for PV powered homes for our own philosophical reasons. We are in practice and by education environmentalists. Each additional PV home is one less home creating a demand for that next oil, coal, or nuclear power plant. It is one more conservation-minded household.

Fowler Solar Electric Inc.
226 Huntington Road
Worthington, MA 01098

* *National Electrical Code®* and *NEC®* are Registered Trademarks of the National Fire Protection Association, Inc., Quincy, MA 02269

Chapter 1

An Introduction to Solar Electricity

In early 1981 Jeffrey Fowler, the founder of Fowler Solar Electric Inc., chose a new project: to design and build an 1800 square foot passive solar house. By May he had sold his properties and moved across town to nine acres of south sloping land. Himself, his trusty little dog Rolo, and Bob and Karin Cook (his neighbors one quarter mile up the road) were the only residents of this 2500 acre tract of land. This area remains undeveloped largely because the last utility pole is one and one third miles, and $20,000, away.

This area is a good wind site. For six months Jeff studied turbines and towers. He also browsed through the photovoltaic information distributed by Joel Davidson in his early ventures. The PV literature went to the waste basket shortly after his estimate for kilowatt usage dictated a $20,000 system.

Bob Cook reinitiated his interest in photovoltaics when he loaned him a federal publication which approached system size from a user standpoint. A case study for each individual PV home indicated the system size and the average appliance use supported by this system. The people cited in this publication were filling moderate needs with much smaller systems than one would have estimated. They had rearranged their use patterns for conservation. Within three months, Jeffrey Fowler and the Cooks had each happily left behind the fume-filled age of darkness for a two module 12V photovoltaic system. These modules were purchased from a Joel Davidson cooperative buy.

After twelve years of research into the components and design of alternative energy systems, we've come to believe that not everyone should engage in the same cumbersome process. This book is designed to be used by the owner or potential owner of a PV system. Much of the product information has been selected and digested to specifically apply to the remote site home owner. We discuss the batteries, chargers, and inverters necessary in our application. Hopefully you will find it more valuable to read about the batteries you can use in your remote site home rather than the batteries available for all PV applications. When applicable, the product information for the most commonly used components has been included.

The PV systems we have designed have sold themselves. The owners are happily committed to their alternative sources of energy. This book proposes several system designs. The logic behind this kit technology is that you, the consumer, will not make the same expensive mistakes others have. A wrong guess or a little misinformation can result in a wrong choice or a poor design. If you purchase the wrong piece of equipment, you forever live with poor performance, or suffer the economic loss incurred in a trade-in. The solution is to profit from the previous mistakes of others.

Photo 1-1 Fowler Home (courtesy of Fowler Solar Electric Inc.)

There are many reasons why a remote site home should be powered by photovoltaic electricity. For the majority of the people, the first reason to use PV is cost. If your home is $10,000 from the power line you are required to invest, or borrow and then invest, the full amount up front. You are responsible for an extra $1000 of cost to run a cable from the road to the house and install the service equipment. Then you have the privilege of purchasing power at the going rate, with guaranteed yearly increases.

PV electricity costs much more per kilowatt than the grid, about two to two and a half times as much even after you account for the full life of all the components. But often this is the choice between a Cadillac and a Volkswagen. If you can happily decrease your usage by the same percentage increase in cost per kilowatt with a home designed for efficiency, then you save the large cost of a line extension to your remote home.

A PV system is modular. It can be purchased first as a beginning system and then expanded to a medium or large system. In the 1980's, economically we were encouraged to buy now and pay later. With this process, we paid later, plus a large amount of interest. If you are not rich, the best way to beat the system is to borrow less and avoid the long term interest costs. Specifically for PV, buy a small system and add to it as your cash flow permits. Another approach is to use a short term loan to take advantage of the state tax credits still available in a few states such as here in Massachusetts. If you have $3000 in November for a PV system, borrow $2000 additional and buy a $5000 system. In March or April, when you receive the tax credits, pay off the loan.

A PV system is superior to a wind system in the home application. There is less maintenance, and what there is does not require an owner to choose between climbing a tower in an ice storm and watching his or her investment go down the drain. The sun shines more frequently than the wind blows. Long periods of no power in a wind system require a large and expensive battery bank. During the last few years most alternative energy enthusiasts have taken down their windmills and gone into the PV business. Wind machines sold today are small and are used as a supplement to a PV system in the winter, when the sun does not shine and the wind blows.

A PV system is a welcome replacement to a life with a generator. Generators are inefficient in cost per kilowatt because they must run constantly at full RPM even if only a light bulb or a TV is being powered. They are noisy and require a lot of maintenance. They are much more expensive than a PV system if length of life and periodic maintenance are realistically considered. If a generator is needed, it should be used in conjunction with a PV system to create a PV-generator hybrid. This increases the generator's efficiency while at the same time decreasing the running time and thus extending the generator's life.

A remote site home is usually sited where it is because the owner wants privacy. Any owner who pays the price for power from the grid will automatically get some neighbors in the process. Contrastingly a person can purchase a piece of land inaccessible to power, have his or her privacy, and pay less because it is not attractive to the majority of people who feel that grid power is a necessity. The lesser price of the land can pay for a large PV system. For the complete recluse, there will be no monthly visit from the nosy meterman.

Some people should not own a PV system. These are the people who expect that it will magically get installed, will never need to be monitored, and will supply enough electricity under all conditions as the power company does. With the power company, you just pay more when the relatives come for a week and leave all the lights on. With a PV system you will conserve or start the backup generator. There are, of course, very large installed, turnkey systems for those with the economic means.

More and more we are selling PV systems to people in the Northeast who are building a new home that is well within reach of the power line. These people are ignoring the economic reality that a PV system will produce electricity that will cost them two to three times per kilowatt what the power company charges. They are often making a political statement against the nuclear power plants or for saving the environment. Some just plain want to do everything independently of the system. The recent instability of the oil market is making these people look more clever, as oil surcharges on electric bills are making the cost per kilowatt of electricity increase dramatically.

Owning and installing a PV system is a significant commitment. You will need to understand its use just as you had to learn to drive a car. You will also need to troubleshoot the basics of the system to be able to call your dealer or the manufacturer. Hopefully, this book will give a greater understanding of a PV system in the remote site home.

In theory, the most basic PV system is composed of a PV module and an appliance. When the sun shines, the module produces electricity and the appliance operates using the electricity. In Figure 1-1, the sun shines on the PV module and the module makes electricity that lights the bulb.

The simple PV system contains a PV module, a battery, and a load. (See Figure 1-2). The module transforms light energy to low voltage DC electricity that is stored in the battery until the electrical load is activated. The load uses up the energy stored in the battery and the PV module then recharges the battery. The PV module could directly power the load. A 50 watt PV module could power a 50 watt light bulb, but only in the daytime, and only on a sunny day. The battery storage can power the load at a time when the PV module is not producing electricity. On a sunny day a 50 watt PV module produces 50 watt-hr per hour for six hours, or 300 watt-hr per day. In this case, the module could replenish the battery after the battery has been depleted 300 watt-hrs per day. A possible load might be three 50 watt bulbs run for two hours each.

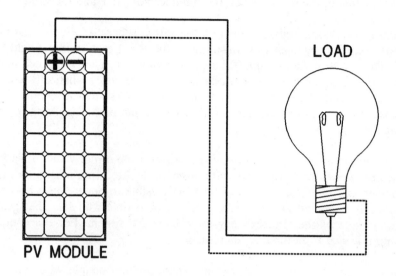

Figure 1-1 A PV Module and a Load

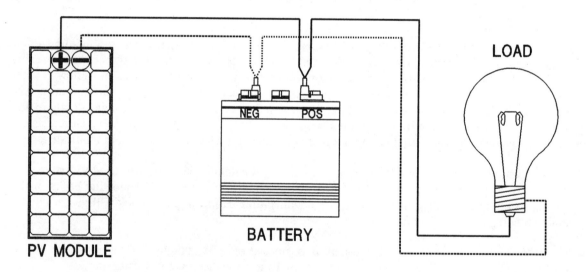

Figure 1-2 A Simple PV System

A PV system does not run on a scheduled daily basis. The sun may shine for two days and produce 600 watt-hrs and then not produce any electricity on a third cloudy day. The load can also be distributed unequally. No load might be used for two days, then three days of load might all be used on the third day. A PV system produces an average amount of electricity dependent on the average amount of sunshine. When a PV module produces an average of 100 watt-hrs per day for ten days, it stores 1000 watt-hrs in the battery. In this same ten day period no more than 1000 watt-hrs are available to power the loads.

A remote site home system consists of a PV array (one or more PV modules wired together), a charge controller, a battery bank, a DC fuse box, an inverter, and a 120V AC circuit breaker box. (See Figure 1-3). Functionally, the PV array and charge controller together are no more than a simple battery charger that uses sunlight as its energy source.

The PV array produces electricity when the sun shines. The charge controller regulates the flow from the array to the battery bank. When the battery bank is low, the charge controller feeds all of the electricity from the array to the batteries. As the batteries approach a state of full charge, the charge controller tapers the supply of electricity to prevent overcharging of the battery. At night it prevents a reverse flow of current from the batteries to the array. The battery bank stores the electricity as low voltage DC, normally at 12V or 24V. The electricity is distributed through a DC fuse box to power low voltage DC appliances. The batteries supply electricity to a device called an inverter which changes the low voltage DC to 120V alternating current and then sends it to an AC circuit breaker box.

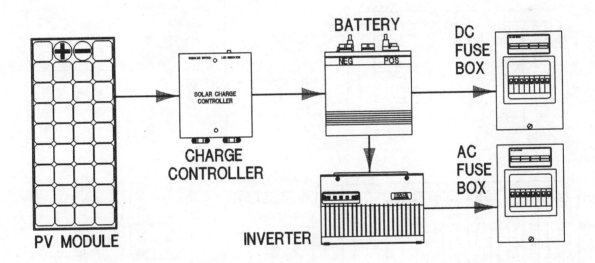

Figure 1-3 Flow Diagram of PV System Components

The charge controller and inverter are purchased in a specific voltage and a specific size based on the proposed performance. The batteries and PV modules, however, are modular. A PV array may consist of one PV module at 12V, or 10 modules producing 10 times the current, but still at 12 volts. Likewise, a battery bank may be enlarged by increasing the number of batteries to allow more storage at the same 12V. A small system can be enlarged at any time by simply adding more modules and more batteries. (See Figure 1-4).

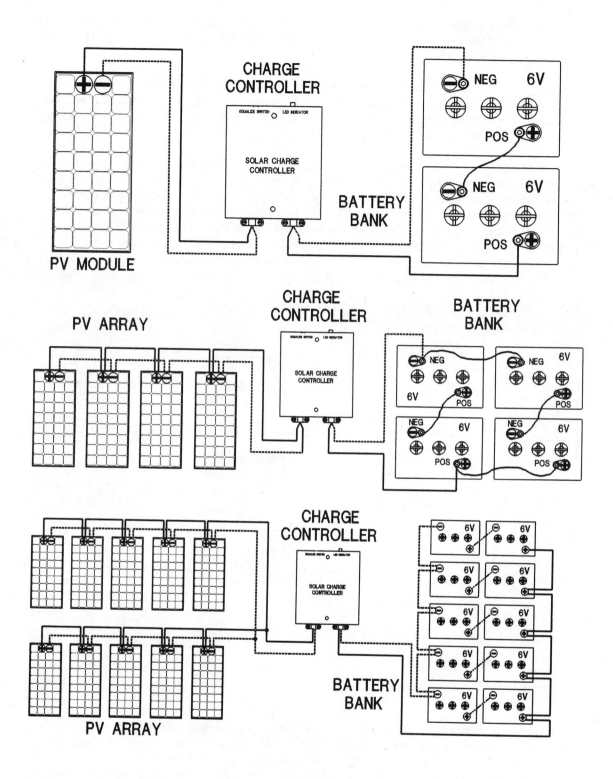

Figure 1-4 Modular Expansion of a PV System

At this point we strongly suggest that you look ahead to Chapter 16, "Understanding the Electricity for Your PV System". This chapter reviews the basic electrical theory that you will need in this book. If you are new to basic electrical theory, please read Chapter 16 before you proceed to Chapter 2. You may also refer to Chapter 16 or the Appendix as you read the other chapters.

Chapter 2

PV Modules and Array Mounting

The single best component in the PV system is the PV module. The world may wait for the technology of production to advance and the price to fall, but the world will never complain about the reliability of the PV module. Modules almost never malfunction. Some are abused, some are subjected to catastrophic events, but it is almost unheard of for one to cease to function of its own accord. In general, manufacturers guarantee their modules to produce no less than 90% of their rated value for a term of ten years. The actual test evidence estimates a greater than twenty year life span.

The PV module is composed of many cells which are wired together in series to create one module which normally operates at 12V DC and a wattage of 30 to 60 watts. (See Figure 2-1.) These cells are laminated between a front piece of impact resistant glass and a back layer and sealer of plastic. This whole structure is in turn framed with anodized aluminum to create a mountable unit. Modules normally weigh ten or twelve pounds and cover an area of approximately four square feet.

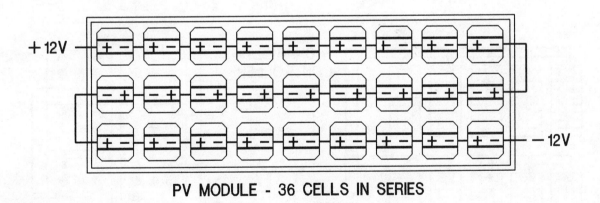

PV MODULE - 36 CELLS IN SERIES

Figure 2-1 PV Module with 36 Cells in Series

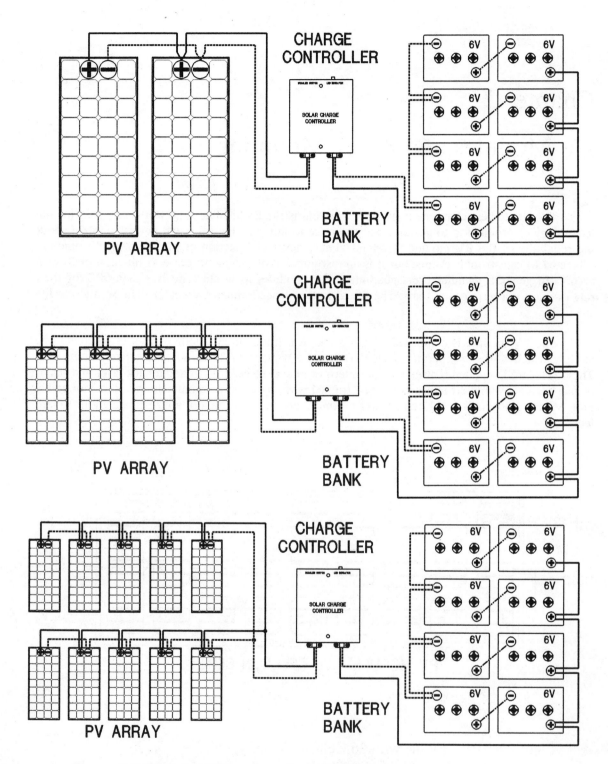

Figure 2-2 Modular Expansion of a 12V PV System

Four of the best known manufacturers of PV modules are Siemens Solar Industries Inc. (formerly ARCO Solar® Inc.*), Solarex® Corporation**, Hoxan America, Inc., and Kyocera International Inc.. All these companies produce good quality modules. One must choose on the basis of price, availability, and size.

Early models of PV modules utilized questionable, exposed terminals. All connections were made on one post with little protection from weather or physical damage. Now single or multiple connections are enclosed in one or two weather-tight junction boxes. Some companies include a bypass diode. These are only really necessary in a system over 24 volt. The PV systems for remote homes will be at 12 volts or 24 volts.

A bypass diode allows an alternate path for electricity when the normal path is not in use. In our case, two 12 volt modules are wired in series to create a 24 volt system. When normal insolation occurs each module produces its 12 volts which additively becomes 24 volts. If, however, one is shaded, then the sunlit module will try to drive the shaded one, which functions as a resistor. At 24 volts or less there is no damage. The voltage is not high enough to damage the shaded module's cells. Some companies wire a bypass diode into each module to allow an alternate path for electricity around the shaded module. This is for applications where many modules are wired in series to create higher voltage arrays.

Potential shading must be considered when siting a PV array. Simply stated, you will need to have no shading of the array during the optimum six hours of midday sun. In the worst case scenario, a tree trunk that casts a shadow diagonally across four PV modules will reduce the output by more than 50%. Many people think that the winter sun low in the sky and passing through tree branches without leaves is fine. It is not. It only gives you 60 or 70%. Of course, this is the time of year when you need it most.

The modular design of a remote site home PV system makes it ideal for a home owner who wishes to start small and expand. A system can consist of two modules which are wired in parallel to produce 12V. This same system could be increased in size with two more modules all wired in parallel, thus still producing 12V. The four modules will still produce 12 volt DC electricity but twice as much. If you added more modules in parallel, you would then have proportionately more power. (See Figure 2-2.)

If the original system of two modules is rewired so that the two 12 volt modules are now in series, then it will produce 24 volt DC electricity. This will be the same amount of electricity but at the new voltage. It too can be expanded but by paralleling additional series wired pairs. (See Figure 2-3.) Batteries can also be added and rewired to change the size and voltage design of the system. An owner is not confined to one system size or design, and not forced to pay the price of trade-in.

Though the different types of cells yield modules of comparable quality, they do create modules with different efficiencies. When we analyze the individual efficiencies in the apples for apples of dollars per watt, we find the market determines a consistent price. However, as a cell's efficiency decreases, it produces less electricity per area; or it takes more area to produce the same electricity. For example, a Solarex® MSX-48 module is 20 inches by 44 inches covering an area of 6.72 square feet and produces 48 watts. At the other extreme, a Hoxan 4810 module is 16.5 inches by 37 inches covering an area of 4.2 sq. ft. and produces the same 48 watts. (See Figure 2-4.) In most cases, area is limited, and a module is desirable that produces similar wattage utilizing a smaller area.

There are good quality lower output modules than the standard 48 to 60 watt modules. The

* ARCO Solar® is a registered trademark of ARCO Solar Inc., Camarillo, CA 93010

**Solarex® Corporation is a registered trademark of Solarex® Corporation, Rockville, MD

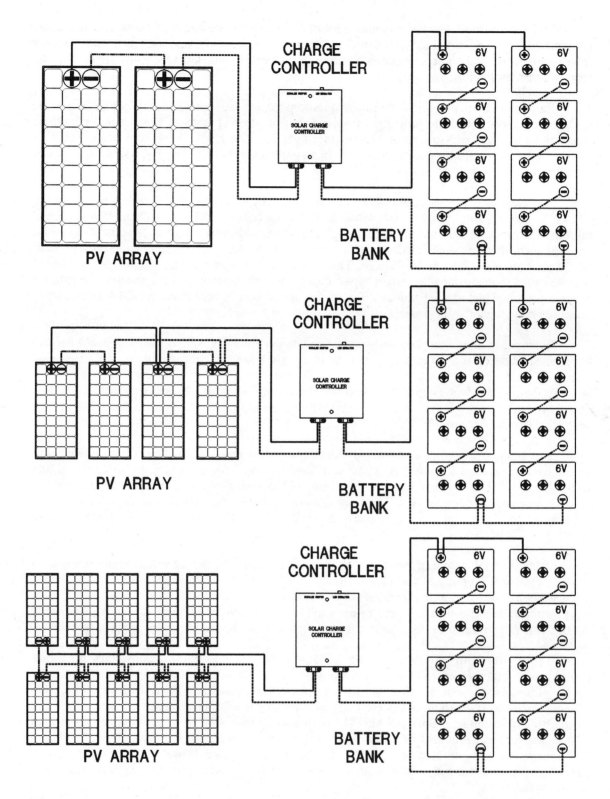

Figure 2-3 Modular Expansion of a 24V PV System

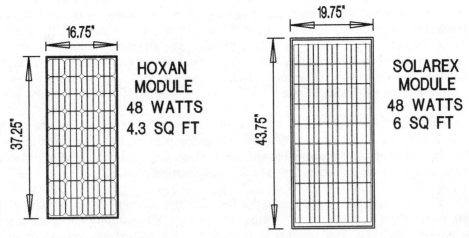

Figure 2-4 Module Surface Area vs. Module Output

disadvantage is that you will need more modules to produce the same power. This is a bad deal, if you are not getting your frames for free. You are increasing the total cost of the array because you are buying more frames now, or in the future, for the same electricity produced.

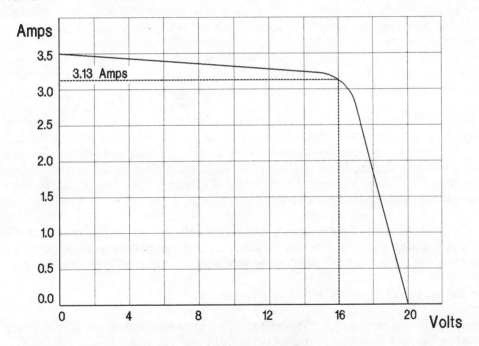

Figure 2-5 Medium Voltage 50 Watt Module IV Curve

When comparing the specifications of several modules, one must be sure that the same standard conditions are used. This should be 1000 watts per square meter at 25 degrees C. cell operating temperature. Another tool for comparison is the IV curve. This is a graph which compares current (I) and voltage (V). If one finds a specific voltage on this curve, it determines a specific current on the other axis. In our application the voltage is determined by the voltage of the battery. This is not just the nominal 12 volts of a 12V bank, but the 13.5 volts to 15 volts of a battery that charges to full charge. Usually it is best to take the voltage at 16 volts and find the current value at this point when you are using a standard cell operating temperature of 25 degrees C. The usable voltage of 16 volts and the usable current of perhaps 3.13 amps multiplied together give the usable wattage of 50 watts. (See Figure 2-5.)

Trying to find out the usable output of a PV module, the price per usable watt, and the best module for your home has always been confusing. We have spent many hours explaining the principles to people. Unfortunately, most people and most sellers only go by the peak watt ratings. We will now try to give you some guidelines that are not mathematical models.

The peak watts of a module is found by multiplying the amps at the peak power point by the volts at the peak power point (amps X volts = watts). The modules you will buy will come in three peak power point voltages, 17V, 16V, or 14.5V. A 17V module is normally composed of 36, approximately, .5V solar cells wired in series. A 16V module is composed of 32 or 33 of the same cells wired in series. And the 14.5V module is composed of 30 cells wired in series. To get the total module voltage you are actually adding up all of the voltages of the individual cells in series. Modules with more cells have higher voltages.

If you live in a hot climate with intense summer sun, such as the sunbelt or south of the USA to the equator, then you need a 17V module. The hotter climate will make the cell temperature of the module go up and it will operate at a lower voltage of around 16 volts.

If you live in a moderate climate, you will only need a 16V module because the module will not operate as hotly and not drop down in voltage. It will give you the 16V needed to charge your batteries. You may also use a 17V module, though the higher voltage is not necessary.

The third, 14.5V, module does not have enough voltage for normal battery charging. As the battery voltage rises as a battery becomes fully charged, the module starts putting out less and less amps. This is not efficient. However, these modules, by default, gradually put less and less into the batteries as the batteries become fuller and fuller. These modules can be used in a simple system, replacing the need for a charge controller if the right size battery is chosen. These modules are normally referred to as self-regulating modules.

The most expensive part of a PV module is the solar cells. If you live in a moderate climate, you can use a 16V module. This module will have less cells. If the module maker charges less for less cells, then you will get the same amps to charge your batteries for less dollars and less space.

When you are comparing modules and output vs. price:

1. Determine if you need only a 17V module. If so, then compare all the 17V modules you are considering by the peak power point amps. Compare the cost per amp.

2. If you can use either a 17V module or a 16V module, compare all the modules in question (either voltage), by the peak power point amps. Compare the cost per amp. FUDGE FACTOR: The 17V module operating in a climate where only a 16V is necessary will actually produce 5-6% more peak power point amps at 16V than it is rated at 17V. Example: A 17V module that puts out 3.0 amps at 17V will put out 3.15 amps at 16V. You should use 3.15 amps for you comparison peak power point amps.

H-4810
Photovoltaic Module

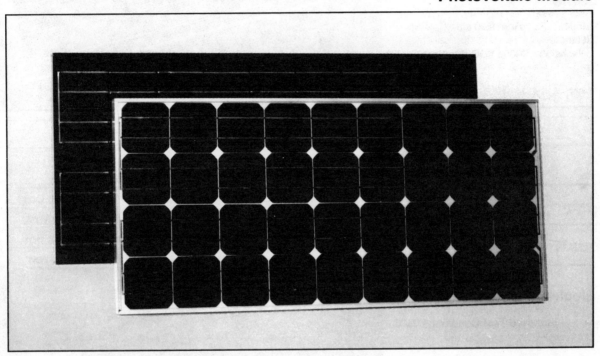

Electrical Specifications

Standard Test Conditions 25°C

Short Curcuit Current I_{sc}	3.2 Amps
Open Circuit Voltage V_{oc}	21.3 Volts
Current at Peak Power Point I_m	2.8 Amps
Voltage at Peak Power Point V_m	17.0 Volts
Power Output (peak)	48.0 Watts

Normal Operating Conditions 48°C

Short Circuit Current I_{sc}	3.2 Amps
Open Circuit Voltage V_{oc}	19.8 Volts
Current at Peak Power Point I_m	2.8 Amps
Voltage at Peak Power Point V_m	15.8 Volts
Power Output (peak)	44.7 Watts

The H-4810 is covered by a ten-year limited warranty on power output.
(See warranty certificate for details)
SPECIFICATIONS SUBJECT TO CHANGE WITHOUT NOTICE

Mechanical Specifications

Length:	945 mm (37 1/8 in.)
Width:	422 mm (16 1/2 in.)
Thickness:	30 mm (1 3/16 in.) Frame only
	45 mm (1 3/4 in.) With junction box
Weight:	6.1 kg (13.4 lb.)
Front Cover:	Low iron tempered glass
Encapsulant:	Ethylene Vinyl Acetate
Solar Cells:	100 x 100 mm square cells; 36 in series
Edge Sealant:	Butyl rubber
Frame:	Anodized structural aluminum
Electrical Isolation:	3000 VDC, 10 microA (typ.)

Siemens Solar

M55

High Efficiency Solar Electric Module

The M55 has 36 cells in series and is well suited for most solar electric applications, including battery charging in the very hottest climates and direct connection to DC motors.

53 Watts* Listed **10 Yr.** Limited Warranty* on Power Output

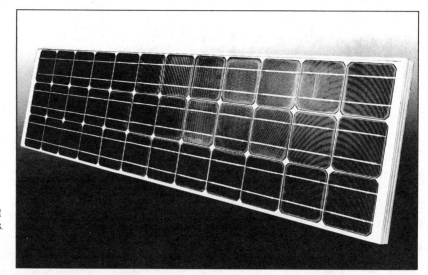

M75

High Efficiency Solar Electric Module

The M75, with 33 cells in series, represents the optimum module configuration for battery charging in all but the hottest climates where the extra voltage of the M55 may be needed. The M75 can also be used for direct connection in many DC motor applications.

48 Watts* Listed **10 Yr.** Limited Warranty* on Power Output

M65

Self-Regulating High Efficiency Solar Electric Module

The M65 is a self-regulating module with 30 cells in series and is designed for direct battery connection. It is intended to meet the needs of those who require reliable electrical power for many applications, particularly lighting, appliances and other equipment in remote homes, recreational vehicles and boats.

43 Watts* Listed **10 Yr.** Limited Warranty* on Power Output

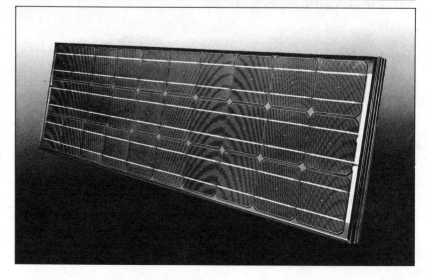

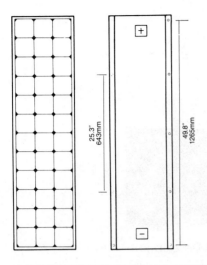

Power Specifications*

Model	M55
Rated Power	**53 Watts**
Current (typical at load)	3.05 Amps
Voltage (typical at load)	17.4 Volts
Short Circuit Current (typical)	3.4 Amps
Open Circuit Voltage (typical)	21.7 Volts

Physical Characteristics

Length	50.9 in / 1293 mm
Width	13 in / 330 mm
Depth	1.4 in / 36 mm
Weight	12.6 lb / 5.7 kg

Performance Characteristics

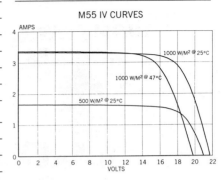

The IV curve (current vs. voltage) above demonstrates typical power response to various light levels at 25°C and a 47°C cell temperature.

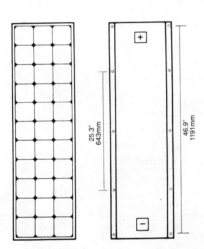

Power Specifications*

Model	M75
Rated Power	**48 Watts**
Current (typical at load)	3.02 Amps
Voltage (typical at load)	15.9 Volts
Short Circuit Current (typical)	3.4 Amps
Open Circuit Voltage (typical)	19.8 Volts

Physical Characteristics

Length	48 in / 1219 mm
Width	13 in / 330 mm
Depth	1.4 in / 36 mm
Weight	11.6 lb / 5.2 kg

Performance Characteristics

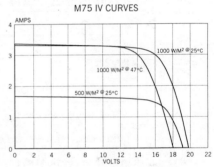

The IV curve (current vs. voltage) above demonstrates typical power response to various light levels at 25°C and a 47°C cell temperature.

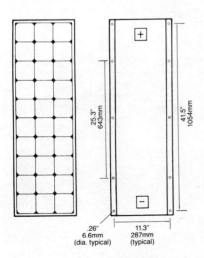

Power Specifications*

Model	M65
Rated Power	**43 Watts**
Current (typical at load)	2.95 Amps
Voltage (typical at load)	14.6 Volts
Short Circuit Current (typical)	3.3 Amps
Open Circuit Voltage (typical)	18.0 Volts

Physical Characteristics

Length	42.6 in / 1083 mm
Width	13 in / 330 mm
Depth	1.4 in / 36 mm
Weight	10.5 lb / 4.8 kg

*Power specifications are at standard test conditions of: 1000 W/M² solar irradiance, 25°C cell temperature and solar spectral irradiance per ASTM E892

Performance Characteristics

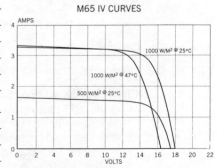

The IV curve (current vs. voltage) above demonstrates typical power response to various light levels at 25°C and a 47°C cell temperature.

HIGH-POWER MODULES

SOLAREX
An Amoco Company

MSX-60

Power Specifications*		Physical Description	
Typical Peak Power**	60W	Length	43.65 in./110.9 cm
Guaranteed Min. Peak Power	58W	Width	19.76 in./50.2 cm
Current @ Peak Power Voltage (17.8V)	3.37A	Depth	2.13 in./5.41 cm
Current @ Operating Voltage (15V)	3.56A	Weight	15.9 lbs./7.2 kg

MSX-56

Power Specifications*		Physical Description	
Typical Peak Power**	56W	Length	43.65 in./110.9 cm
Guaranteed Min. Peak Power	54W	Width	19.76 in./50.2 cm
Current @ Peak Power Voltage (17.8V)	3.16A	Depth	2.13 in./5.41 cm
Current @ Operating Voltage (15V)	3.37A	Weight	15.9 lbs./7.2 kg

MSX-53

Power Specifications*		Physical Description	
Typical Peak Power**	53W	Length	43.65 in./110.9 cm
Guaranteed Min. Peak Power	50W	Width	19.76 in./50.2 cm
Current @ Peak Power Voltage (17.8V)	3.0A	Depth	2.13 in./5.41 cm
Current @ Operating Voltage (15V)	3.22A	Weight	15.9 lbs./7.2 kg

OPTIONS (MSX-60, 56, 53)
- Protective aluminum backplate
- Mounting hardware kits
- Blocking and/or bypass diodes
- Solarstate™ voltage regulator
- Marine environment junction box
- 6-volt output

---

*Power specifications are for standard 12-volt shipping configuration
**Peak power is defined as the maximum amount of power available from the module under Standard Test Conditions (STC) which are:
— Illumination of 1 kW/meter^2 (1 sun) at spectral distribution of AM 1.5
— Cell temperature of 25°C

After you have considered all the technical specifications of the available modules, you must face the fact that the PV array will probably be very visible. Modules come in different shapes and thus fit in different spaces. They have distinctive designs and colors. If architecture as well as performance is important, then choose accordingly.

Photo 2-1 Tracker (Courtesy of Home Power Magazine - Brian Green)

ARRAY MOUNTING

Modules may be mounted on a pole, a ground support, a side wall of a building, or a roof. The first consideration obviously is day long access to unobstructed sunlight; but also, it is important to consider wire length to batteries and just plain where they look good.

Anytime a module's orientation is exactly perpendicular to the sunlight it receives the greatest possible insolation and puts out its highest wattage. Ideally, a mount that can change directions so that it will always be perpendicular to the sun will have the greatest output. In theory, this would mean the array would follow the sun from east to west and also tilt upward at the same time as the sun rises higher in the sky at noon.

An array mounting structure that changes its orientation so that it will stay perpendicular to the rays of the sun as the sun moves across the sky, is called a tracker. Most trackers on the market only follow the sun from east to west. In parts of the country blessed with consistently high insolation, a tracker is an excellent investment. The price of the tracker increases the total output of the array for less money than it would cost to buy more modules and to get the increased output.

Photo 2-2 La Fontaine Home (Courtesy of Fowler Solar Electric Inc.)

In higher latitudes than the sunbelt the tracker works great in the summer months but not well in the winter months. In the winter the sun rises in the southeast and sets in the southwest. The sun never has an orientation way to the east or way to the west. The angle from where the sun rises to where the sun sets is small. Thus the array never has a very bad angle to the sun and a tracker cannot improve the insolation much; consequently it cannot increase the output much. The tracker would pay for itself in the summer only. Unfortunately, this only helps when there is an increased summer load or if the home is only used in the summer. Year round homes traditionally have their greatest demand for electricity in the winter when lights are on longer. In the north, money is better invested in additional modules that will increase winter output.

Modules that point due south are in the best position to be most perpendicular to the sun's light from 10 AM to 2 PM when the sun is at its highest intensity. Modules do not, however, need to be pointing due south. It is more important to point them within 25 degrees east or west of true south if the day-long access to sunlight is enhanced this way. For example, if a mountain to the right shadows the module at 2 PM and a valley to the east allows morning sunlight, then face the modules more easterly. Experiment and plot out with the aid of solar charts, where the sun will be in other seasons. Relate this to the trees and mountains specific to your site.

At noon, sunlight is at its highest intensity. Modules facing due south are in the best position to capture this sunlight at the optimum time. The sun is also highest in the sky at this point, thus a module must also be directed to the position of the sun in terms of height above the horizon. Unfortunately for module mounts, the sun's noon elevation changes 47 degrees from December to June. A standard fixed mount is positioned at an angle from the ground that is the latitude of the area plus 15 degrees. This assumes you are north of the equator. At a latitude of 44 degrees, the modules would be mounted at an angle of 59 degrees. This orientation gets the best winter exposure and the best snow reflection.

The fixed mount is a good alternative to varying the mounting angle. There is no maintenance and the array is installed and done with. The system is increased in size to compensate for its compromised efficiency. This works well for a system of 10 modules or greater, or a cabin system used only on the weekend. For year round use, the angle of the fixed mount is best at latitude plus 15 degrees. This maximizes the winter gain. The summer reduced output is compensated by the increased number of sunny days and the decreased lighting loads.

Most home owners do not size their systems for worst case winter use. They size their system on what they can afford, and what they will be able to additionally afford in the next few years. Then they cut back their use to fit the season. These home owners will want optimum efficiency from the system. The solution is to use a mounting design that allows seasonal adjustment. Normally, two adjustments per year are adequate. In northern areas of winter snow cover, a December or January adjustment to a sun low in the sky will pick up a precious 20% of insolation from snow reflection. A spring-fall adjustment can give enough additional insolation to power a washing machine or DC refrigerator. Rarely does a summer adjustment increase the output enough to go from two adjustments a year to four adjustments. Decreased labor or material costs for a frame that has only a winter adjustment can be put into more modules.

Many module manufacturers sell mounting frames to fit their modules. These do not adjust and must be ordered for specific angle orientation. One solution is to order standard mounts and modify them so that they can be adjusted. These frames are aluminum, and they are pre-drilled to match the modules. They require no long hunts for materials. One disadvantage to manufactured mounts is that most companies mount the modules with the four foot direction east-west and stack the four or six modules. (See Figure 2-6.) In snow country, the modules should have the four foot direction going up. Snow slides better off one span of four feet rather than a series of one foot spans with frame edges in between. Some larger PV businesses, such as ourselves, manufacture their own frames to satisfy this situation. (See Figure 2-7.)

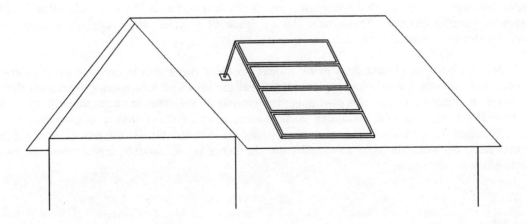

Figure 2-6 Standard Mounting Structure - Modules Oriented Horizontally

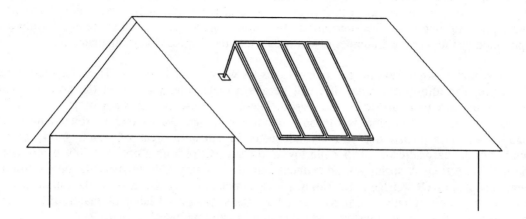

Figure 2-7 Snow Country Mounting Structures - Modules Oriented Vertically

Module frames, as previously mentioned, are made of aluminum. Aluminum, when in contact with a dissimilar metal such as iron or steel, sets up an electrolytic reaction that causes oxidation. Aluminum panels must be attached to aluminum frames only with aluminum or stainless steel nuts and bolts.

Home owner designed frames can be made from iron or steel, but there must be an insulating separator, such as a rubber washer or a stainless steel washer between the module and the frame. Stainless steel bolts can be used as fasteners because they are compatible with both iron and aluminum. Wood can also be used if the frame is designed to last, and withstand wind and weather. Modules are very rigid; often a full frame is not needed but only a rail connecting each set of ends of parallel modules. These rails can be made of recycled angle iron (bed springs) or galvanized electrical conduit.

Modules must be grounded. A metal mount on the ground should be attached to a driven ground rod. A wooden mount must have modules that are all wired to a good ground such that, if one module is removed, it does not disconnect the ground of any other module. Metal frames need only have the frame grounded assuming each module is so attached that it is grounded to the mount. If a fixed mount is chosen, it must always have good ventilation at the rear of the modules. Modules must dissipate the heat absorbed from the sunlight. As module temperature increases, module efficiency decreases.

Chapter 3

Charge Controllers

The charge controller in a PV system is a voltage regulation device which matches and protects both the PV array and the battery bank. If the modules were allowed to constantly charge a battery bank, then in times of low load, they would be capable of supplying more wattage to the batteries than the batteries could hold. Simply, more electrical energy would be going into the battery bank than would be drawn out. A deep cycling lead acid battery, which is normally used, would reach a state of overcharge. In a lesser case, the batteries would boil off too much water and require extra maintenance. In a severe case, the batteries would be damaged.

A controller's job is to sense the level of charge of the battery bank by measuring its voltage. When the battery reaches a desired high voltage, all or part of the PV current must then be channeled away from the battery. When only part of the current is shunted, the remainder functions as a maintenance trickle charging current to keep the batteries at 100% charge.

The second function of a charge controller is to connect the modules to the battery bank in the morning and disconnect them at dark. During sunlight conditions the modules produce a current which charges the batteries. However, at night when no current is being produced, the batteries leak or discharge backward through the PV array. This can cause inefficiency in some cases and module damage in others, depending on the design of the system.

There are four or five designs for charge controlling in PV systems. A remote site home will use a series, a shunt, or a series-shunt charge controller. These names describe the internal circuitry of the units. The controller may utilize a conventional circuit board or it may be a sealed solid state unit. The latter is used in an environment where it will be subjected to extremes in temperature or weather conditions. Since a fully solid state unit does not have a relay, it is advisable in areas where sparks or arcing are a hazard. Often these units utilize a blocking diode to prevent reverse current flow to the array at night. The diode consumes a small portion of the module current flowing to the battery, lowering its efficiency. In low voltage systems, the blocking may really be unnecessary because the amount the array leaks at night is about the same as the diode consumes in the daytime.

A typical controller senses the voltage being supplied to it by the PV array. After dawn, this voltage rises above the level of the battery bank and the controller closes a relay completing a circuit from the modules to the battery. The full amount of the current being produced by the modules flows to the battery bank. This mode will continue until dark when the controller senses a low voltage from the non-productive modules, and opens the relay, disconnecting the modules from the batteries, and thus eliminating any leakage backwards from the batteries to the modules.

Every day, or on any given day, the modules will be supplying current at a rate that surpasses the amount the batteries have previously been discharged. In other words, the modules will fully charge the batteries. As the batteries go from low charge to full charge, their voltage steadily increases. The controller senses the high voltage level and disconnects the PV modules from the batteries preventing overcharging. The bulk of the disconnected current is dissipated while a small percentage is still supplied as a trickle maintenance charge. This level of charging cannot overcharge the battery bank.

If a load is connected during the day which exceeds the wattage of the trickle charge mode, then the battery bank will supply the balance of electricity. This will discharge the battery and will pull the voltage down indicating that the battery is less than 100% charged. When the controller senses this lower voltage level, the controller will reconnect full current from the PV array. If a load is never connected, then the trickle charge continues until disconnect at dark. The next morning full array current is reconnected.

In a typical 12V PV system, the array will continue to charge until the batteries reach 14.7 volts. The full charge mode will disconnect and a 1 or 2 amp trickle charge will maintain a voltage of 13.5V. When a load is applied which pulls down the voltage to 12.5V or below, the full charge mode will be reconnected. For a 24V system, all values are doubled. The above description describes the workings of the SCI Charger™* Model 1 or Model 2 controller.

A recent charge controller on the market is the Sun Selector®**. The modules charge the batteries until they reach 14 volts then they are disconnected. When the batteries dissipate their surface charge and drop down to 13.2 volts, the modules are reconnected. This causes a pulse of full charge then an off time. As the batteries become fully charged, the off time is longer and the on time is shorter, thus creating a simulated trickle charge. When the on time becomes very short the controller overshoots the high voltage point. This results in a short voltage spike every time before it pulses to off. This spike simulates an equalizing charge for the batteries. Thus, we have a controller that equalizes but gasses the batteries less, making them safer and extending their life by using them less hard.

Trace Engineering Co. has recently upgraded its C-30 controller to a C-30A. The original unit simply turned the array on when the voltage was low and off when the voltage was high. A diode was necessary to prevent nighttime backfeeding. There were no indicator lights or meters.

The new C-30A has a circuit to prevent backfeeding to the modules that is more efficient than a blocking diode. It has an LED readout that monitors the controller's functions. There is a manual override switch for battery equalization. This of course simply connects the array directly to the batteries no matter what the voltage is. You can keep the modules connected to drive the voltage above the normal disconnect point to do a planned overcharge.

This C-30A has field settable voltage points. We suggest you set them to 13.2 reconnect and 14.4 disconnect. This means the controller will work like the Sun Selector® we have described above. The batteries will be equalized under most conditions as long as they have not been discharged for a long time. The batteries can be tested periodically to see if they are equalized. If they are equalized, all cells in the bank will have a specific gravity reading that is within 20 points. If not, use the override. To equalize the unequal cells, the fully charged cells are overcharged while the lower cells are gradually catching up. At the end of the process all of the cells with have an equal and full charge specific gravity on your hydrometer.

There are no meters on the Trace C-30A. Though the LED readout is helpful, we suggest

*SCI Charger™ is a trademark of Specialty Concepts Inc., Canoga Park, CA 91304

**Sun Selector® is a registered trademark of Bobier Electronics Inc., Parkersburg, WV 26101

you add an add-on meter package.

Many options are available on controllers.

METERS

One voltmeter and one ammeter are imperative in any system. It is easiest to order these as a factory option and get meters that are matched to the size of the unit. Most meters will only be standard analog meters that are not very accurate. This will still be fine. The ammeter need only indicate that the modules are producing approximately the right amps. The voltmeter can be calibrated with a digital voltmeter upon installation.

LOAD DISCONNECT

Small DC loads pass through the controller before distribution. If the battery discharges too low, the controller automatically disconnects all loads. The loads are not reconnected until the batteries are charged again to a 75% level. Good system maintenance in your own home is far better than low voltage disconnect. If you have a weekend cabin, or if you travel and are forgetful, you may need this option. A normal good sized system, powering a forgotten light, will never trip this device anyway. For good wiring, DC loads should be powered from the batteries directly whenever possible. If a low voltage disconnect is a must, we recommend using a Trace C-30 load disconnect separate from the charge controller.

TEMPERATURE COMPENSATION

Battery storage capacity and full charge voltage level are dependent on temperature. The charge controller with a temperature compensation option self-adjusts its voltage disconnect points according to temperature. Obviously if the sensor is internal to the charge controller, then the controller must be installed in a location that is the same temperature as the batteries. This option is normally considered when the battery bank is less than 20 degrees F. The best solution is to find a better place for the batteries. You may also select a charge controller that has field settable voltage points. You may then set the voltage points higher for the winter.

ARRAY DIVERSION

When batteries reach a full charge the array current is switched to another load. Only under special cases can this be used in a remote site home system. The array current can be switched to another battery bank, but you will need an additional charge controller to keep the second bank from over charging. Any other option to use the diverted current must be able to take the full range of amperage and voltage from the array, few appliances will not be damaged in this situation.

In a normal remote site home, a basic series or shunt controller with meters is the basic advisable controller. Controllers are rated for an array of a specific voltage and a maximum amperage. For example, a controller may charge a 24V battery bank from an array with a maximum 30 amp output. To reach 30 amps at 24 volts, an array could be composed of ten pairs of Hoxan 12V, 48 watt modules, each pair wired in series with a maximum output of 3.0 amps at 24 volts per pair. Or a controller may charge a 12V battery bank at a maximum 30 amps. This might serve an array up to ten Solarex® MSX-48, 12V modules in parallel at a maximum output of 2.85 amps per module. Of course, a lesser number of modules would work on either controller. If a larger array is needed, a controller can be paralleled with another identical controller, thus doubling the maximum total array size. (See Figure 3-1.)

In northern climates with winter snow cover, the array will increase its peak amperage output by up to 25%. This means that an array that just stays under 30 amps in the summer will overproduce in the winter and run over 30 amps. The controller may fail in this condition. Check with your dealer for advice on the specific unit.

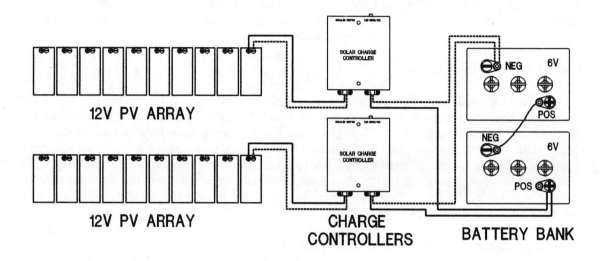

Figure 3-1 Two Arrays and Charge Controllers Charging One Battery Bank

Chapter 4

Batteries

The most difficult assignment in a book of this nature is to include the necessary information, but not too much information. For simplification we will concentrate on lead antimony, lead acid, deep cycling batteries. At the current level of battery production and technology, this is the one good choice for remote site home installation. These batteries are, of course, not ideal for PV use, but the economy of scale dictates a price that compensates for their less than perfect design.

There are superior designs of batteries such as the Absolyte®* suspended electrolyte batteries. The life is longer, the temperature range of operation is better, and the maintenance is negligible. However, the price is prohibitive even after factoring in the increased life span. The battery industry should develop a battery specifically for home PV systems. Until your first battery bank replacement, use the deep cycle battery.

Maintenance-free batteries are available for PV systems. Some are sealed with liquid electrolyte, some are captive electrolyte, and some are gel cell. This sounds great. They need no addition of water and will not gas. However, there are problems. You must actually be more careful with them than standard deep cycle batteries.

These batteries will not gas, only if they are charged at no more than their maximum recommended charging voltage. If this is surpassed, the battery may be damaged. The least damage will be that the batteries will gas and the gas will pass out of a pressure release vent. Unfortunately, you will have no way to later add the lost water that was broken down to create the gas.

You will be able to custom set a charge controller to match the setpoints needed for maintenance-free batteries. The problem occurs if you ever have a charge controller that fails, and does not disconnect the modules from the batteries. If you are away, then you could lose the whole expensive battery bank.

Most maintenance-free batteries need to be returned to full charge in a few days after they are deeply discharged. The greatest problem with batteries in PV systems is when more electricity is taken from the batteries, in a time like winter, than is put back into them. The batteries do not reach full charge again for several weeks or a month. Most maintenance-free batteries are not meant for this use. The standard deep cycle battery is much more tolerant. When the standard battery is completely abused from lack of recharging, there is still a fix-it solution. The batteries can be overcharged for a period of time. This will result in large water losses, but the water can be replaced.

*Absolyte® is a registered trademark of GNB Inc., Langhorn, PA 19047.

Do not use used batteries, unless of course, they are perfectly matched amongst themselves, are nearly new, and are cheap to cover all your miscalculations. In other words, unless they are perfect. Do not use automotive batteries. They will work but not for long enough for the price paid. Never use a used car battery unless it was free and is in your camp where you don't need any dependability anyway. Never try to mix and match batteries of different sizes, ages and manufacturers. All the best ones will run no better than the worst battery in the whole pile once they are connected in a bank.

The lead antimony, lead acid, deep cycle battery is simply a battery which uses lead plates, improved with the addition of a small percentage of antimony, and sulfuric acid as the electrolyte. Deep cycling is a description of the battery design which is capable of use that will discharge the battery to a twenty percent charge, and then recharge it to full charge. From full charge to low charge to full charge constitutes one deep cycle. An automobile battery contrastingly is designed for shallow cycling and high cold cranking amps. The plates are constructed differently to make a greater surface area available to the electrolyte. Frequent deep cycling will ruin an automobile battery.

Deep cycle batteries can be broken down further into two groups. One group is referred to as electric vehicle, golf cart, marine or forklift batteries. These batteries are available in 6V or 12V sizes. Only consider the 6V size. A 6V battery compared to a 12V battery of the same capacity or storage, or even easier the same weight, has half as many cells and half as many plates, each of which is twice as heavy duty. These heavy plates will last longer. A battery of this type will normally weigh 60-70 pounds and have a 180-220 amp-hr rating at a 20 hour discharge rate.

The second group is industrial deep cycle batteries. To further muddy the waters some of this group are referred to as marine or forklift batteries. These batteries may come in 2V, 6V or 8V. They are normally available in large sizes. An individual battery may weigh 120 pounds to 300 pounds. These are heavier duty batteries and longer lived. Often, they are not applicable because the twelve large 2V batteries that are minimum to create a 24V battery bank form an oversized and over expensive storage bank. These batteries are manufactured manually on a special order rather than mass production basis. They are more expensive per amp-hr.

Most systems will use a 6V electric vehicle battery with a 200 amp-hr rating based on a 20 hour discharge rate. In simplest terms, this means one can discharge a full battery and receive 200 amp-hrs at 6V or 1200 watt-hrs over a 20 hour period. Do not be confused by the twenty hour discharge rate. The important thing is to compare batteries at the same standard conditions. When comparing amp-hrs, make sure each has the same hour rate. Normally this will be 20 hours, but sometimes 24 hours.

To compare different manufacturers' batteries, first compare amp-hour ratings and price at a constant 6V 20 hour discharge rate. Next, when possible, choose the battery by the number of cycles it has available in its average life. These cycles will be defined as 80% discharge or 50% discharge. Get the same rating for each battery. If the number of cycles is not available, remember that this is largely determined (once you've sought out a good quality battery) by battery separator material. Plate thickness, if unavailable, is indicated by gross weight of the battery. If you are considering two different 6V 200 amp hour batteries and one weighs 65 lbs. and one weighs 72 lbs., the second will be a heavier duty model, depending on case design and weight. Try to get a battery with glass mats and rubber separators.

The last important specification of a battery is plate composition. For our batteries it will be the percentage of antimony. Your battery dealer will not know this.

Battery dealers are not engineers. We have to ask the right questions just the right way from the right factory engineer to get good answers. Your best solution is to seek the advice of your most knowledgeable PV dealer and buy the batteries from him or her. He or she should have experience besides specifications. We would recommend that you avoid situations where a PV dealer or battery dealer advertises a model battery or its equivalent. An equivalent battery means amp-hr rating, not necessarily length of life.

Whenever possible you should purchase batteries that are fully charged with the acid in them. If you think they have been sitting for a long time, test them for specific gravity for assurance that you will not need to charge them and that the plates are not sulfated. Dry charged batteries will be shipped without acid. You will need to add the acid to each cell and, worse still, they will need to be charged. This will take too long with a generator and charger or, even worse, with your PV system. You must charge them before you install them at the site.

For many years we primarily sold the Exide®* GC-4 battery. A GC-4 weighs 65 lbs., wet, and is rated at 185 amp-hours at 20 hour discharge rate. It should be noted that these ratings vary according to how optimistic the manufacturer is. This model Exide® is conservatively rated. It will compare in size with most companies' 200-220 amp-hr golf cart batteries. It has glass mats and rubber separators and a life expectancy of 500 cycles at an 80% discharge per cycle. We are happy with these because they are well designed and well made. The oldest ones installed are now over six years old. We have not had a cell fail for any reason whatsoever. Exide® also supplies an EE-88 model which sells for ten or fifteen dollars less. It has a proportionately lower amp-hr rating for its price. However, it does not have glass mats and rubber separators, and consequently has two thirds of the life. The true dollar value is in the best battery.

Beware of the stories that the lesser battery is just as good. We have had customers that had other batteries and have had to tell them the bad news when one of the batteries in a bank failed after only two years. The bad news was they might as well replace the whole bank.

Currently, we primarily sell the Trojan®** T-105 battery. This is a 64 pound 6V battery rated at 217 amp-hr at a 20 hr. discharge rate. The Trojan® batteries have a new patented battery separator that tests show results in longer battery life. The T-105 is rated at 630 80% discharge cycles. This translates into a 20% longer life than the Exide® GC-4.

We also sell the Trojan® L-16 battery. This is a 130 pound 6V battery that is rated at 350 amp-hr at the 20 hr. discharge rate. This is one of the most popular batteries in PV systems. The plate design is heavier than the T-105.

We are constantly asked by customers whether they should buy a battery bank of T-105's or L-16's. This is how we advise them. In many ways it is a toss up as far as dollars go. The L-16 is estimated by Trojan® to last around 900 cycles. The T-105 has been tested to last 630 cycles. The L-16 should last 30% longer, but it will cost you 35% more per amp-hr.

The advantage of the L-16 is fewer batteries for the same size battery bank and fewer battery interconnections. The first disadvantage is external cell bus bars. These create more exposed electrical terminals and a faster self-discharge rate. Spilled electrolyte from battery testing gets under these interconnect bars and is hard to clean. The second disadvantage is weight. A 130 pound battery is awkward for two people and very heavy for one person. If a heavy battery is dropped even a few inches, the case may break or a plate may break inside. The T-105 has a more modern case

*Exide® is a registered trademark of Exide® Corporation, Horsham, PA 19044.

**Trojan® is a registered trademark of Trojan® Battery Company, Santa Fe Springs, CA.

that has internal cell interconnections. It is easy to handle. The general rule of thumb is, consider the L-16 for battery banks of eight or more L-16's. However the T-105's may still be your choice.

If your PV dealer is too far away in terms of shipping, and if no batteries are manufactured or distributed near you, then go to your local Sears store. Theirs will only be average batteries but they will be locally available. We recommend that you use the largest 6V deep cycle model, not the 12V marine model. The 12V marine battery is really more of a heavy duty automotive battery than a true deep cycle battery.

A deep cycle battery has, by design, heavier plates than an automotive battery, which has thinner plates and more surface area to supply the heavy cold cranking amps needed to start a cold motor. We do not recommend 12V deep cycle batteries in the 60 pound and under size because the plates are too thin and they cannot be a true deep cycle battery. You need a set number of plates in any 12V battery. When the battery is small, the plates must all be thin. If, however, the 12V battery were 120 pounds then each individual plate would be twice as thick. This, unfortunately, makes a heavy battery that is difficult to move and install. There is a greater likelihood of dropping the battery and damaging the inside plates.

The solution is to get two 6V deep cycle batteries that each weigh 60 or more pounds. They are easy to install and once the battery cable is installed between the two, you have the equivalent of a 12V battery. The 12V battery has six cells all in one large case. The only difference here is that you have one 6V battery of three cells that is externally connected to the next three cells of the second 6V battery.

A lead acid battery is most efficient when it is discharged at a slow rate. This is ideal for a properly sized PV system. Most loads are small and are on for a long period of time. Even a 1000 watt pump or a skill saw operated for five or ten minutes discharges the battery bank at a slow rate. A good battery may have a life expectancy of 600 full cycles. A well designed PV system should only be discharging the battery bank one fifth of its

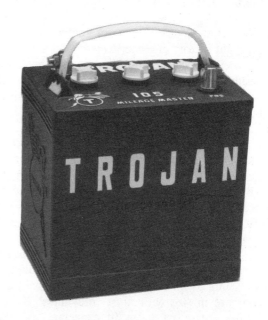

Photo 4-1 Trojan® T-105 (Courtesy of Trojan® Battery Company)

Photo 4-2 Trojan® L-16 (Courtesy of Trojan® Battery Company)

storage capacity in one day. This is a 20% cycle per day or one full cycle every five days. If the life expectancy of the battery bank is 600 full cycles, then the battery bank has a life expectancy of 3000 days or eight years.

There are standard procedures for testing the number of cycles of a deep cycle battery. The battery is discharged at a 75 amp rate until the battery is down 80%. Then the battery is recharged quickly by an industrial charger. In our PV systems we use battery much less harshly. They may do better than the industry ratings when used in a PV system.

An average deep cycle battery at full charge and at rest (this means it has not been receiving charge for a few hours and is not under load) has a specific gravity of 1.260 and a voltage reading of 12.6V. As it is discharged, the specific gravity and voltage fall to 1.120 and 11.85V at complete discharge. When the battery is again charged, it will accept a very high charge rate which must be gradually tapered as the battery reaches 75 to 100% charge. In a good system with a good controller, the batteries will be charged to 14.5V before a maintenance trickle charge is used to hold a 100% charge. A full charge condition is necessary every 4 weeks to prevent the batteries from losing part of their full storage capacity.

The sun does not shine every day. If, for instance, it does not shine for four days and the batteries are drained to 80% in the four days, this is still a 20% cycle per day. For good battery life, a deep cycle battery should not be discharged below a 20% charge. A battery will normally survive a complete discharge but the more often this happens the shorter its life will be. It should never be discharged more than 80% at a high rate or with a heavy load. A light load would be 200 watts of light bulbs and a heavy load would be a 1000 watt skill saw. Some inverters have a low voltage disconnect to prevent such battery damage.

Lead acid batteries perform ideally at 77 degrees F. They perform well at 35 to 45 degrees F. in the winter in a basement. The lower temperature only causes a 10% to 15% decrease in storage capacity. At higher than 77 degrees Fahrenheit, they perform better but at a decreased life. If batteries must operate over a greater temperature range, temperature compensation may be necessary on the charge controller. For most remote homes, we worry when the batteries are in a cold environment below 32 degrees F.

Many people in the PV industry advise bypassing the charge controller periodically to overcharge the batteries to equalize all cells in the batteries. We have systems that have been running on a standard SCI Charger™ Model (2) controller for six years with no equalizing overcharge and have all cells in the system equal. The idea of equalizing charge comes from the battery industry where they charge with industrial chargers at high rates in a short period of time. In a PV system, we have low charge rates which means the voltage at which the cells equalize is lower. Normally the batteries are driven up to 14.5 volts at noon at the highest charge rate. However, in the afternoon the modules may only be supplying half the charge rate and still sending the batteries up to 14.5 volts. At the lower charge rate, this 14.5 volts is an equalizing charge. There are conflicting theories on this. Practically it works.

It is important to make sure to get the batteries back up to complete charge every 4-6 weeks. Don't leave them discharged for long periods of time. If the batteries do not periodically return to full charge, then the coating of sulfate will crystallize on the surface of the plates. This impedes the normal charging process of the battery. The battery will effectively store less electricity. To remedy this situation the battery must be overcharged to drive the sulfate coating off the plates.

The above condition where the batteries are not returned to full charge for a long period of time contributes to unequal cell charge levels. It is good practice to test all cells every three months or any time you know they have not reached full charge in several weeks. If the cells are unequal then manually overcharge until the lesser cells come up to the specific gravity of the fully charged cells.

<u>Batteries are dangerous. They should have their own area that is well ventilated. As they approach full charge, electrolysis splits water molecules into hydrogen and oxygen molecules. In a confined space these only need a spark from a DC motor or charge controller relay to explode.</u>

CAREFULLY HEED ALL BATTERY MANUFACTURER WARNINGS. BATTERIES CAN ARC AND BURN OR EVEN EXPLODE. THE ACID INSIDE WILL BURN SKIN AND BLIND EYES.

If sixteen 6V 200 amp-hr batteries are confined in a sealed 8 foot by 8 foot room, there is an extreme problem. If however, this same room is ventilated to the rest of a normal, not very airtight cellar, there is little danger. Still, under gassing conditions, sparks should be kept away from the tops of batteries. A good and simple system uses a polyethylene hood which hangs above the batteries (not down around them) and is gathered to a one inch pipe that vents outside. We recommend ventilation of your battery bank. Consult the local electrical code about battery safety.

In the past, the only way to prevent gassing of batteries was to purchase a much more expensive battery of a different technology. Now we recommend using lower setpoints on charge controllers to reduce gassing. The modules charge the batteries at full amperage until the batteries reach 14.4 volts. The modules are disconnected and remain so until the surface charge of the batteries dissipates and the battery voltage drops below 13.2 volts. The modules are then reconnected. As the batteries become fully charged, the off time is longer and the on time is shorter. Any controller with adjustable set points can copy this. We use the Trace controller with adjustable setpoints. The result is a controller that reduces gassing greatly and extends the life of the batteries.

The Trace C-30A controller has a manual override switch that stops the controller from disconnecting the modules. When the switch is in this mode, the modules can overcharge the batteries. We feel it is better to plan and manually overcharge and gas your batteries rather than to having them doing it every day.

Batteries are most dangerous when someone drops a metal tool and short circuits two terminals. Sparks and fire result. A battery that completely shorts can overheat and melt and explode spraying acid everywhere!!! Beware!

Photo 4-3 Battery Bank (Courtesy of Fowler Solar Electric Inc.)

CHAPTER 5

INVERTERS

The inverter is potentially the weakest link in the remote site home PV system. It is a complicated, expensive piece of electronic technology, which to most owners comes in the shape of a mysterious large black box. Inverters for PV homes have come a long way in the last seven years. The best inverters are very reliable. They will require less warranty service than a generator. The majority of them will never do anything but run. Most PV homes will either immediately or in a short while need an inverter.

The PV system, in the simplest analysis, is a battery bank and a battery charger. Modules and batteries produce and store DC electricity. DC current moves in one direction. In most remote site homes, it is supplied at either 12V or 24V. The conventional house connected to a utility line runs its appliances at 120V AC. Alternating current moves first in one direction, then stops its flow and then flows in the opposite direction. This sequence constitutes one cycle. The change in flow is not obvious because there are sixty of these cycles occurring each second. Thus, utility current is 120V AC at 60Hz (cycles/sec.).

An inverter is an electronic device that takes an input of DC electricity and produces an output of AC electricity. In low voltage PV systems, the inverter transforms the low voltage electricity to 120V electricity. The first job the inverter does is switch the current on and off while it also keeps reversing its flow. This produces an alternating current which is still at the original low voltage. This is then put through a step-up transformer that yields 120V AC electricity. The switching must also be performed at a precise speed to produce 60 Hz. The abrupt switching of a simple inverter produces a square wave, which is actually a graphic representation of how the current is flowing. (See Figure 5-1.) Utility current flows in a sinusoidal wave. Its current is switched on and gradually increases until it reaches a peak, then gradually decreases to off, and gradually increases in the negative direction. (See Figure 5-2.)

Inverters which are capable of producing true sine waves are not cost effective for remote site homes. Square wave inverters are too inferior for any use except resistive loads which produce light or heat. Instead, most stand alone inverters approximate a sine wave and refer to it as a modified square wave or a quasi-sine wave. In contrast to a square wave, a modified square wave changes the width and height of the square wave, depending on the load, such that the average is close to a sine wave. For non-electrical engineers it is best to imagine taking the square wave and replacing it with a staircase that is not a sine wave but much closer than the basic square wave. The more steps, the closer the approximation. (See Figure 5-3.) The stand alone inverter is an imitator of utility electricity. The better its imitation, the higher its quality. It must produce a voltage that is accurate over its whole range of output, a precise 60 Hz, and a good waveform.

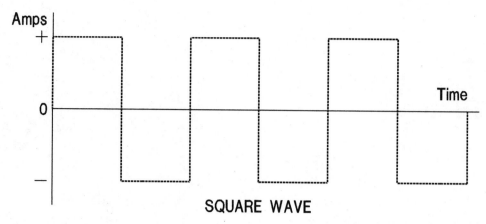

Figure 5-1 A Square Wave

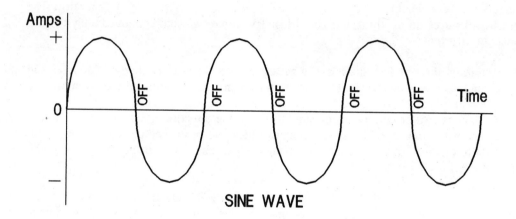

Figure 5-2 A Sine Wave

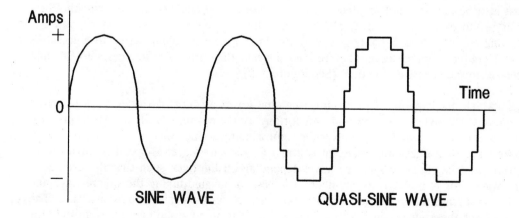

Figure 5-3 Comparison of a Sine Wave and a Quasi-sine Wave

Power factor is a rating of the synchronization of current waveform to voltage waveform. When a stand alone inverter powers a light bulb, it merely needs to produce a steady voltage and current. Because of the design of an inductive motor, the current waveform and voltage waveform are not the same. This is a reactive load. If an inverter is powering a reactive load and a non-reactive load, the reactive load changes the waveform of the inverter, resulting in an interference in the other appliance. In practice, a washing machine causes a light bulb to flicker or strobe. A desirable inverter will compensate for reactive loads and maintain a good waveform.

A 100 watt inductive motor draws 300 watts momentarily and then 100 watts continually. A basic inverter must be sized, not for the motor usage, but for the initial surge. Therefore, a 300 watt inverter would be needed to start the 100 watt motor. Inverters are rated for their continuous power output. A good quality one will normally rate its surge ability. This is a problem, however, because the ratings are not consistent. It is helpful to get a list of appliances an inverter will run. A good inverter should withstand an instantaneous surge of three to four times its continuous output and a short time surge of one and a half to two times its continuous output.

Inverters offer many features. Some are optional and some are standard equipment.

LOAD DEMAND

This is necessary on all remote site home systems. An inverter that is turned on and is supplying no load still consumes 8-40 watts. Load demand has a circuit that waits for a load, consuming only one to four watts. When a load is applied, it senses it and turns on the main inverter.

OVERLOAD PROTECTION

If the inverter is overloaded by a large appliance or a short, it will shut down. This is normally done electronically. The unit turns itself off. It is also done thermally for loads just over the inverters limit. The inverter becomes hot running a large load too long.

DIRECT WIRING

In addition to the receptacles, the inverter can be wired directly to a service panel.

D.C. CIRCUIT BREAKERS

The circuit breaker disconnects the inverter from the battery bank. If one is not supplied with the inverter, some sort of other fusing must be wired between the battery and the inverter.

REMOTE SWITCH

Additional wire and switch are supplied to allow an inverter to be turned on and off from another location.

STANDBY OPTION

The inverter is equipped with a large battery charger to be powered by a generator for back-up charging of the batteries. This option includes an automatic transfer switch that connects all loads powered by the inverter, straight through to the generator whenever the generator is running.

STACKING OPTION

An interface for electrically grouping two inverters so they will function as one double sized inverter.

THERMAL COOLING

This is a thermal sensor and a fan to cool the inverter or battery charger when it produces a lot of heat during sustained heavy loads or heavy charging. It improves the inverter's efficiency and keeps it from disconnecting from over temperature.

DIGITAL MULTIMETER

A metering system, that is internal to the unit, monitors battery charging from a generator when the inverter has a standby option.

LOW VOLTAGE DISCONNECT

Many inverters have a low battery disconnect, but it is set to protect the inverter if the batteries are dead. This optional low voltage disconnect has adjustable setpoints to shut off when the batteries are low but far from dead.

240 VOLT TRANSFORMER

This option is a transformer to transform 120V AC output of the inverter to 240V AC output and 120V AC output. This is most needed for a 240V AC submersible pump.

An important consideration in any part of a remote site home PV system is efficiency. Again, there is no standard for measurement. Companies will claim efficiencies from 80 to 90%. The true efficiency of an inverter is the output wattage in AC divided by the battery input in DC. If an 80 watt AC appliance is powered by an inverter that is using 100 watts DC, then there is an 80% efficiency. For general purposes a good stand alone inverter will yield a 60-80 percent efficiency when total loads are less than 50 watts, and 80-90% efficiency for loads above 50 watts. Most inverters require a minimum amount of wattage to be on. If 40 watts is used, and a 40 watt bulb is turned on, then the inverter uses 80 watts to run a 40 watt bulb, for a 50% efficiency.

The Trace inverters use 1/3 of a watt to stay in standby mode, and 8 watts when the unit is on but powering no loads. This makes even a small AC load such as an efficient 15 watt Osram light, an efficient use of the inverter. This light will only use 25 watts to light a whole room after inverter inefficiencies.

Not all inverters produce the same waveform, or the same waveform the same way. Inverters cause interference in electronics when their waveform is not clean enough. A heavy filtering circuit in an inverter could eliminate interference, but it would also greatly reduce efficiency. The solution is to get an inverter with a small amount of interference but still a high efficiency. AM radios will pick up intolerable interference when the inverter is powering a load. New electronic telephones will pick up a little background noise, especially if you are silly enough to run the wires in the walls through the same holes with the house wiring. A bad practice anywhere.

Televisions, stereos, and computers normally have switching power supplies. There is a lot of evidence that inverters do a better job of running this kind of power supply than regular AC. People always ask us about running computers. We have been running them for years here. The only problem seems to be that if a customer has a computer fail, the store instantly tells him or her that it must be the inverter. In actuality, small computer UPS systems use an inverter, but one that is cheaper and of poorer quality than the one you will use. We do recommend the use of a good surge protector. Please remember that your computer is being powered by an inverter with only so much power to give. If you turn on another load that is too large, you can pull the inverter voltage down and crash the computer. Here we use IBM clones with hefty power supplies; even the water pump and washing machine coming on only blips the screen, but does not crash the computer.

Televisions will sometimes exhibit a small interference on the screen. But unless it is an old model, you will only be able to notice it when you know where to look for it. VCRs and microwaves also work fine. Stereos must be tried. Older tube models will hum badly as will some new very expensive units. However, most recent $200 to $500 receivers or cassette decks will run just like they do at Mom's. If you are buying a new appliance for your alternative energy home, always make sure you can return it after you have tried it at home. Never tell them about your power system. The salesman will not understand and instantly be sure it will fail or tell you he won't warranty it. Ignorance is a crime.

The inverter is potentially the weakest link in the PV system, but this is only relative. In fact, this is the weakest link compared to a system which is so simple and maintenance free that no other part may ever fail. Until inverters become completely reliable, consider the ease of repair into the criteria for choosing an inverter. Choose an inverter that is serviced or manufactured locally, or is light weight enough to travel UPS instead of freight. Most local TV repair people will not be able to trouble shoot an inverter. Nothing is worse than to find out it will cost $90 and take three weeks to receive the transcontinental free warranty service.

Heart Interface was the most popular inverter for remote home use. It was the first company to address the needs of this market. Their inverters reached a peak efficiency at a very low load level, one or two lights. This meant some houses could go all AC and utilize efficient lights such as the compact fluorescents. They ran motors efficiently and they had good surge capabilities. Their deficiencies were reliability, changing internal adjustments, and slow warranty repair; shipping to the factory on the larger units required slow and expensive land freight. The inverter had many potentiometers on the circuit board that were cheap and finicky to adjust. The adjustments we suspected changed in shipping. Many of these questions have now been addressed by Heart Interface.

Probably the most popular inverter today is the Trace. The Trace Engineering Co. is a spin-off of Heart. Trace has solved the problems in the Hearts. Their units all travel UPS. This means cheaply and quickly with the option for overnight delivery. This shipping means a cheaper delivered price and a quicker and cheaper warranty repair. So far we've had few warranty repairs. They boast a two year warranty. The unit has no adjustments to fail. The inverter is monitored by itself and changes its internal workings as its temperature changes and correspondingly its parts specifications change. The basic 2000 watt 12V model or the 2500 watt 24V model will run the right 1/3 HP deep well pump and the right washing machine at the same time. These inverters have very high surge capabilities and will run large loads for the short periods of time that are needed in a remote site home. The threshold turn on level is programmable with DIP switches and the standby battery charge models offer a completely programmable heavy duty battery charger. Turn on voltage, turn off voltage, and charging amperage are all user selectable. The best part of all is that

Specifications

Rated Power @ 20 deg. C.	Model 612 — 600 watts Model 2012 — 2000 watts Model 2524 — 2500 watts Model 2232 — 2100 watts Model 2236 — 2200 watts Model 2248 — 2200 watts
Surge Power @ 20 deg. C.	Model 612 — 1800 watts Model 2012 — 6000 watts Model 2524 — 6200 watts Model 2232 — 6000 watts Model 2236 — 6200 watts Model 2248 — 6200 watts
No Load Current	Model 612 — .025 amps Model 2012 — .030 amps Model 2524 — .018 amps Model 2232 — .016 amps Model 2236 — .025 amps Model 2248 — .022 amps
Input Voltage	Model 612 — 9.0 to 15.8 Model 2012 — 9.0 to 15.8 Model 2524 — 19.5 to 30.7 Model 2232 — 20.0 to 42.7 Model 2236 — 30.1 to 47.0 Model 2248 — 39.0 to 61.6
Load Sensing (watts)	Programmable @ 2, 6, 16, 40 or defeated
Voltage Regulation	117 VAC plus or minus 2 volts
Frequency Regulation	Crystal controlled, plus or minus .04%
Power Factor	1 to 1 all conditions allowed
Wave Form	Modified Sine wave with dynamic impulse phase correction for inductive loads
Protection Circuitry (with auto reset)	High battery, low battery High temperature Overcurrent—instantaneous linear temperature compensation
Protection Circuitry	Reverse grid connection Induced electrostatic charge- transient absorber 6500 amps (8x20us 1 time) Overload shutdown after 15 seconds
Dimensions	2000 Series — Height 6.25", width 10" depth 12.4" 612 — Height 5.75", width 10.5", depth 8.0"
Weight	2000 Series — 38 lbs 612 — 14 lbs

Standby Option

Transfer Relay	1 HP 30 amp (2000 Series)
Maximum Charge Rate	Model 612 — 25 amps Model 2012 — 110 amps Model 2524 — 50 amps Model 2232 — 39 amps Model 2236 — 35 amps Model 2248 — 25 amps

Performance Graphs

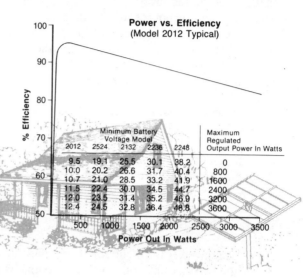

Power vs. Efficiency
(Model 2012 Typical)

Minimum Battery Voltage Model					Maximum Regulated Output Power In Watts
2012	2524	2132	2236	2248	
9.5	19.1	25.5	30.1	38.2	0
10.0	20.2	26.6	31.7	40.4	800
10.7	21.0	28.5	33.2	41.9	1600
11.5	22.4	30.0	34.5	44.7	2400
12.0	23.5	31.4	35.2	46.9	3200
12.4	24.5	32.8	36.4	48.8	3600

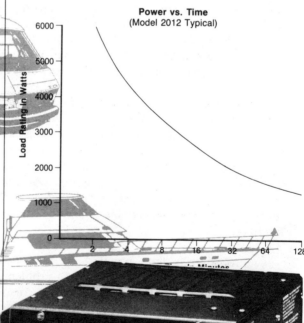

Power vs. Time
(Model 2012 Typical)

they are much cheaper than any of the competition.

The last question that needs to be asked about any PV system is, "Is an inverter necessary?" A house will run nicely on low voltage DC if lights, TV and stereo are the only appliances required. Once motors are needed, it becomes more of a problem for someone to become an appliance expert than it does to deal with an inverter. Large loads just aren't feasible at a low voltage. They require a high amperage and large special wiring. The United States is an AC country, and many appliances are not convertible to DC. If an inverter is not presently needed, avoid it, but remember that it will be needed in the future. Try to buy an inverter only once. Pick the right size and the right voltage. It is expensive to trade in a used inverter.

Inverters' specifications vary depending on the model and the manufacturer. Inverters output potentially deadly voltage. Installation and design of this equipment and their systems should be carried out by qualified personnel. This chapter is not an installation guide.

Photo 5-1 Hopkins Home (Courtesy of Fowler Solar Electric Inc.)

Chapter 6

The Independent Home as a Whole

When designing a PV system the remote site home owner is always faced with a choice of whether to use AC or DC electricity. The smallest system which only operates a few lights and a radio is probably best run on 12V DC. Twelve volts is the standard voltage used in the automotive and RV industry. Incandescent bulbs, car stereos, TV, fluorescents, and many appliances are readily available at local stores and from mail order catalogs. The other bonus of the 12V system is the high efficiency. A 50 watt 12V incandescent bulb produces as much light as 100 watt 120V light. This is not magic; a 12V bulb at 50 watts draws five amps, while a 50 watt bulb at 120V draws .4 amps. Tungsten filament designed for larger amperage produces more light and less heat. For a simple test, touch a 12V bulb. It is not as hot; therefore less energy goes to heat and more to light.

Twelve volt systems do not match well with high wattage appliances. The high amperage and low voltage require wires that must be too large, too short, and too expensive. Even for lights the circuits must use 10 or 12 gauge wire with only a few lights per circuit. The batteries must be centrally located to facilitate short wire runs. The modules also produce 12V and must not use a long wire run.

As a system increases in size, AC electricity and an inverter usually become necessary. AC for most systems is synonymous with 120 volts. Though a system can be 120V DC, it will need ten 12V modules in series and ten 12V batteries in series. This system may be feasible, but the next increment of expansion will be ten more modules and ten more batteries. Still, one will need to search for AC-DC appliances. These will not include deep well submersible pumps, washing machines, or electronic equipment.

In new construction, a house should always be wired up to code with AC. Retrofit wiring is enormously expensive. Someday AC wiring will be needed by the present or future owner. Even a direct generator hook-up is a step from the dark ages.

The best of both worlds is a dual voltage PV system with DC circuits for low voltage efficient lighting for a light in each room, and inverter produced AC current for efficient compact fluorescents and intermittent loads such as water pumps, washing machines, and vacuum cleaners. We prefer 24V low voltage because it permits four times as long a wire run as with 12V at an identical wattage. Lighting is now available in 24V from most PV suppliers. Medium sized inverters, 2000-3000 watts, also use 24 volts.

A dual voltage household can be very confusing to the owner, the visitor, and the electrician. Each voltage system must have its own fuse box, circuits, and receptacles. Low voltage systems appear to be harmless because they rarely shock people. They must be fused. They can start a fire with a short more easily than a 120V system. DC electricity will quickly ruin and short out any wall switch or appliance which is rated AC only. Wall switches for DC are the old fashioned style which make a loud click. The two voltages can never share the same junction box or fuse box.

The *National Electrical Code®* requires that a second voltage DC system must use receptacles and plugs which are a different configuration than the AC and of the locking type. The locking style is a nuisance and probably will not be required by an inspector. However, the second plug configuration is essential. Even the owner of a carefully labeled system will absentmindedly ruin an appliance if all the plugs are the same.

The two largest problems for a remote site homeowner are water supply and refrigeration. Water is a necessity in a hygienic and toilet flushing culture. A handful of households have gravity fed systems; and in some parts of the country with no winter, a small DC pump can slowly fill a large storage tank for gravity feeding. The rest of the country gets its water from a well and most of these are deep wells. For the minority who have shallow or surface wells, DC pumps are the best choice. They are available at decent prices from manufacturers of quality pumps. A DC pump is not subject to inverter malfunctions or generator breakdowns. The DC part of a PV system is nearly 100% reliable. A home with a broken inverter will always have water and DC lights.

Flowlight™* Solar Power makes the DC pumps that are most popular in PV systems. They are reasonably priced and long-lived. They have been manufacturing their Slowpump™** for many years. This pump will draw water up a distance of 20 feet at sea level and one foot less for each thousand feet above sea level. The pump is called a Slowpump™ because it is designed to pump a small number of gallons per minute compared to a conventional household pump. It will correspondingly run for a longer period of time. It will also run at a lower wattage. This design results in low amperage load on your PV system spaced over a longer period of time. The low amperage draw is compatible with longer wire runs at the low voltage DC.

Slowpumps™ come in many sizes and voltages. A specific pump model is chosen depending on the specific requirements of your well and your home. These pumps can be used in a standard 30 PSI / 50 PSI pressurized system. They may also be powered directly by a PV array with no batteries, for bulk water storage. Whenever the sun shines, the pump pumps. For water systems where the pump will not need to push the water up a hill or to an elevated storage tank, but only to pressurize the water system, Flowlight™ Solar Power manufactures a faster pumping model called a booster pump. This will pump at an average 5.5 gallons per minute in a standard pressurized system. We recommend Flowlight™ Solar Power pumps for most homes that have shallow wells because they supply the same gallons as conventional pumps for a lot less electricity. Water pumping is a large part of the budget in any PV home. So an efficient pumping system means a better PV system.

Low voltage DC submersible pumps are available for deep wells. Their use eliminates the need for an inverter or the need for an inverter that is large enough to power an AC well pump and an AC washing machine at the same time. Some of these pumps may not be capable of pressurizing a standard pressurized system. If you choose a non-pressurizing model, you will need a storage tank and a pressurizing pump, or a gravity tank system. Line loss resulting from a long wire run at a low voltage is a potential problem with a DC submersible. If your deep well is close to your home, and correspondingly, the distance from the battery bank to the well and down to the pump is not excessive, you may be able to use an efficient DC submersible.

Flowlight™ Solar Power is in the final stages of producing a Hydra-Jack™*** hydraulic

*Flowlight™ is trademarked by Flowlight™ Solar Power, Santa Cruz, NM 87567.

**Slowpump™ is trademarked by Flowlight™ Solar Power, Santa Cruz, NM 87567.

***Hydra-Jack™ is trademarked by Flowlight™ Solar Power, Santa Cruz, NM 87567.

FLOWLIGHT™ BOOSTER PUMP

DC PRESSURE PUMP FOR INDEPENDENT POWER SYSTEMS

Provides "TOWN PRESSURE" for HOME WATER SUPPLY from 12 or 24 Volt power

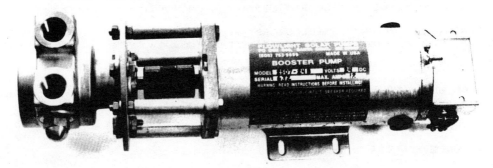

FLOWLIGHT™ PUMPS are the QUIET, EFFICIENT, and RELIABLE way to provide pressurized water from a shallow water source -- well, spring, stream, storage tank or cistern. FLOWLIGHT™ BOOSTER PUMP provides pressure equivalent to an elevated tank 100 feet high! A booster pump system is FAR more cost-effective than an elevated tank, and easier to protect from freezing.

HUNDREDS of FLOWLIGHT™ PUMPS are in use world-wide for remote homes, ranches, vacation lodges, fire lookout towers, campsites and purification systems, and full-time motor homes and RV's...since 1983. They are an industry standard in photovoltaic (solar-electric) applications where efficiency and reliability are vital.

FLOWLIGHT™ BOOSTER PUMP outlasts DOZENS of plastic RV/Marine pumps. It provides MORE FLOW and MORE PRESSURE. Our long-life motor is FOUR TIMES the size. FLOWLIGHT uses a ball bearing, solid brass rotary-vane pump with hardened carbon-graphite and stainless steel moving parts -- no plastic. It's QUIET, producing a steady flow without pulsating and vibrating your pipes.

FLOWLIGHT™ BOOSTER PUMP is self-priming and can draw as high as 20 vertical feet at sea level (subtract 1 ft. for every 1000 ft. altitude). Pressure range is adjustable. A pressure relief valve is built in for safety, and the motor is thermally protected against overload. FLOWLIGHT is a LIFETIME PUMP designed for decades of use. It's fully rebuildable, in case of accidental damage.

WHY DC? Alternative energy and mobile power systems are based on batteries, which store DC power. Conversion to AC to run a conventional pump requires a large, costly power inverter. Installing a DC pump is a SIMPLER, more reliable and economical approach. Our extra-efficient permanent magnet motor and positive displacement pump uses HALF THE ENERGY of an AC pump run by inverter, and requires 1/10 the starting surge!

SPECIFICATIONS

AVERAGE FLOW RATE (at 30-50 PSI): 5.5 GPM

CURRENT DRAW (at 30-50 PSI): 15-25 AMPS (12V model). 7 - 12 Amps (24V model). At lower pressure, current draw is proportionally less.

MAXIMUM CUT-OUT PRESSURE: 50 PSI

DUTY CYCLE: Continuous to 30 PSI, near-continuous at higher pressure.

ENERGY USAGE (typical family home using 165 gallons per day): 10 Amp-Hours/Day at 12 Volts (120 Watt-Hours/Day) -- the average usable output of one typical (40 watt) photovoltaic panel.

LENGTH: 16.5 inches (18.5" with Dry Run Switch) **WEIGHT:** 15 lbs.

FITTINGS, inlet and outlet: 1/2" female pipe thread. Use plastic or brass (no iron) at pump inlet.

PRESSURE CAPACITY: If pump is lifting only a few feet, Max pressure capacity is 50 PSI. For higher vertical lifts (measured from surface of water source to pressure tank) subtract 1 PSI for every 2.3 feet and adjust pressure switch accordingly. Horizontal distances are of little significance if pipe is sized properly.

Specify 12 or 24 Volts.
WARRANTY: 2 YEARS against defects in materials and workmanship

Product Info. 6-1 Flowlight Booster Pump

SYSTEM REQUIREMENTS: PRESSURE TANK (captive air type preferred, any size), PRESSURE SWITCH and PRESSURE GUAGE. These items available from any water supply / pump dealer. ADEQUATE WIRE SIZE: for 12V, min. #10 for up to 12 feet, larger for longer distance. 24V pumps require 1/4 the wire size. Undersized wire will reduce flow rate but will NOT damage the motor.

INSTALLATION: Same as conventional AC pressure system, except for filtration requirement. Pump is smaller and may be installed using flexible pipe. Our COMPLETE INSTRUCTION MANUAL answers all your questions!

WARNINGS: MUST NOT RUN COMPLETELY DRY (dry run protection available). FILTRATION REQUIRED: Super-precision pump must be protected from ALL dirt and fine silt. Use a drinking water filter or one of our filter options.

ACCESSORIES

- **DRY RUN SWITCH:** Shuts off pump if intake runs dry. Specify: MANUAL RESET for water sources that do not reliably recharge (tank, cistern), <u>or</u> AUTO RESET (for slow wells etc. that recover on their own).

- **FINE INTAKE STRAINER/FOOT VALVE:**

- **INLINE SEDIMENT FILTER:** Transparent housing, 10 inch, 5 micron cartridge

- **30" INTAKE FILTER/FOOT VALVE:** Replaces both intake strainer and inline filter with single high capacity submersed cartridge unit. Good for silty streams, drilled well casings, and other problem applications. Spare 30" 5 micron cartridge included. Also accepts 10" cartridges.

- **CLOSE ELBOWS:** Special fittings allow installation within 6" well casing.

- **EASY INSTALLATION KIT:** ALL THE SMALL PARTS you need to quickly install your booster pump system: Accessory Tee, Pressure Switch and Guage, Check Valve, Drain Valve, Shut-Off Valve, Pipe Nipples and Two 18" Flexible Pipes with Unions. ALL BRASS FITTINGS.

OTHER FLOWLIGHT™ SOLAR PUMPS

SLOWPUMP™ is similar to FLOWLIGHT™ BOOSTER PUMP but slower in flow rate, with higher lift capacity. SLOWPUMP™ may run from a battery system or direct from a small photovoltaic array, pumping all day to fill a storage or stock tank. (SLOWPUMP™ may also be used to lift AND pressurize water). 21 different models allow SLOWPUMP™ to precisely fit your requirements, for flow rates from 8 to 250 gallons per hour at lifts as high as 450 feet.

HYDRA-JACK™ pumps slowly and efficiently from any deep well with a casing of 3" or more, for vertical lifts to 400 feet. Similar to jack pumps, but far less expensive and easier to install and service. Sucker rod, rigid pipe and counterweights are eliminated by simple hydraulic coupling using flexible pipe.

ECONO-SUB™: Low power deep well submersible, lifts as much as 230 feet using less than 100 watts of 12 or 24 volt DC power, flow rate from 30 to 100 GPH. Installs easily by hand. A breakthrough in deep well solar pumping.

DEEP WELL INSTALLATIONS: SLOWPUMP™ and BOOSTER PUMP are NOT SUBMERSIBLE, but may be installed in 6" or larger well casings where suction limit will not be exceeded and water will not rise to submerse the pump. Many regions have a sufficiently stable water table to do this. Otherwise, consider HYDRA-JACK™ or ECONO-SUB™.

FLOWLIGHT SOLAR POWER • P.O. BOX 548

SANTA CRUZ, NM   87567 • USA

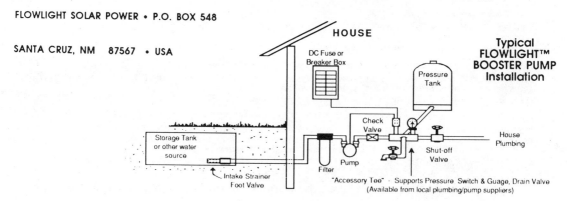

Product Info. 6-2 Flowlight Booster Pump

piston pump. This is a hydraulically powered jack pump. A pump in the basement, or at the top of the well, pumps water down the well to hydraulically move the piston pump at the bottom of the well. The conventional sucker rod and mechanical apparatus are eliminated. We think this has great promise. The DC motor can be near the battery, eliminating the problems of long wire runs at low voltage. Unfortunately, these are quite expensive.

Deeper wells in the dry parts of the country will need specifically designed and expensive pumping systems. These will probably require a DC jack pump or a DC hydraulic piston pump. Please consult your local PV dealer.

For deep wells that have a static water level higher than 150 feet below the pressurized tank in the home, it may be simplest to use a conventional AC submersible pump and an inverter. The AC pump will be less expensive than a DC submersible and the inverter will be used in the PV system for many other appliances.

The shallower of deep wells can use an AC jet pump, but most deep wells use an AC submersible pump. Any pump can push water uphill; its limitations are sucking it up on the intake side. The greater the distance a pump must draw up water, the more inefficient it is. A submersible solves the problem of draw by being submersed in the well. It only needs to push water out of the well. Jet pumps are available, but as the depth of the well increases a jet pump becomes less efficient. The conventional grid house and pump manufacturers are not concerned with wattage used. A remote site homeowner must factor this in.

A submersible pump is underwater somewhere near the bottom of the well. This will be at least 40 feet underground and most likely much farther. A wire leads back up the well and back underground, a distance of up to several hundred feet, to the fuse box in the house. The point is that the motor will be powered by a wire at least a hundred feet and more likely four hundred feet from the power source. This means the 1000-1300 watt submersible pump must either run on high voltage or be powered by mammoth cables to fend off a voltage drop in the line. A conventional submersible pump runs on 240V. These pumps are AC of course.

An AC submersible pump must be powered by a generator or an inverter. A good water system design uses a pressurized system with a 30 psi to 50 psi pressure switch and a 120 gallon galvanized tank. This will provide about 80 gallons of water flow up one floor from the tank between pumpings. In a generator powered system, the generator can pressurize the tank in five minutes and be shut off. Likewise in an inverter system, the inverter does not need to be constantly on.

Two hundred and forty volt power is not standard on inverters. One hundred and twenty volt pumps are available in the one third horsepower size. Only buy a top quality one like Franklin Electric. This will have a maximum current draw of 9 amps. If the 120V line is less than 350 ft., 10 gauge wire is adequate. A deeper well will require a half horsepower 240V pump; or perhaps, a well already has a 240V pump. The 120V power from the inverter or generator can be changed to 240V with a step-up transformer. This will be a heavy duty transformer with a price of $250. This should be wired in the circuit on the pump side of the pressure switch. This way the transformer will only draw power when the pump is running. Any AC submersible pump used with an inverter should be a three wire type with the control box and a starting capacitor in the basement. Two wire pumps may use have a split-phase motor and no starting capacitor. This model may overload the inverter.

In any circumstance with any pump, look up the manufacturer's requirements for wire size vs wire distance (this is from the pump motor to inverter). Don't trust the electrician or the plumber.

Whenever possible use the next size larger wire; this will keep voltage losses to a minimum. A 10% voltage drop, even if acceptable to the motor, is a 10% loss in efficiency.

Well pump people are used to serving the needs of conventional homes. They do not understand that the remote site home will have a more conservative use. Often a 1/3 HP pump will suffice when a regular home would need a 1/2 HP one. An example would be a 400 foot deep well, with a static water level 20 feet deep, that only yielded three gallons per minute. Normally this would require a 1/2 HP pump installed 375 feet down the well. The long wire run would dictate a 240 volt pump. In an alternative energy home, we do not have heavy continual use demands such as watering the lawn all day. We would use a 120 volt 1/3 HP pump which would produce five gallons per minute, and install it 150 feet down. The pump could, of course, pump faster than the well could recover but there are hundreds of gallons above the pump in the well casing which is effectively a storage tank. Our short time heavy use would be compensated by the buffer zone above. Then when the pump was off the well would recover this column of water. (See Figure 6-1.)

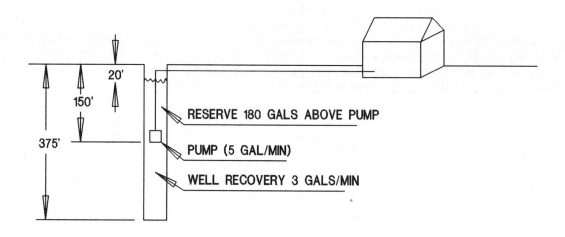

Figure 6-1 An Installation of a Smaller Pump for Conservative Water Needs

A 1000 watt motor requires a surge of 3000 watts to go from its inertia of rest to full speed. A deep well pump always starts under load. The water above it in the pipes and the pressure in the tank are resisting it all the while it is going from rest to full speed. A 1000 watt submersible pump requires a 6000-10,000 watt instantaneous surge. A 2500 watt to 3000 watt generator will absorb the instantaneous surge. An inverter must be rated at 2000-2500 watts with a 150% surge, or a three to four time instantaneous surge, to start a 1000 watt submersible pump.

A standard efficient model household refrigerator uses 400 watts to run its compressor. The compressor is running 50% of the time. This comes to a total of 4800 watt-hrs per day. Even the best quality apartment-size unit in a cool cellar would use 2400 watt-hrs per day. After an electric stove and an electric water heater, the refrigerator is the highest use appliance in a grid house. Though it may only draw 400 watts, it is running 12 hours per day. This appliance is too demanding for the remote site home system and must be replaced.

Like the water heater and stove, refrigeration can be powered by propane. RV distributors carry small gas refrigerators. Beware; many of these require a constant supply of 12V current that will tax your PV system.

A top quality Sibir brand refrigerator from Switzerland was manufactured and distributed in the U.S. The largest model sold for about $1100 depending where you bought it. It contained 8 cu. ft. of efficient space, two of which were a separate door freezer compartment that was a true deep freezer. The Sibir company has recently been purchased by the Dometic Corp. They sell a similar 8 cu. ft. refrigerator that is manufactured in Sweden. This same unit is also sold under the name of Servel™*. Dometic bought the name of the old Servel company. There is no relation.

We no longer recommend that customers look for old Servel refrigerators. Too many are bought in a non-running condition and never do run. When the cooling units weaken or fail, no recharging is available. They are now scarce in New England and thus overpriced. Furthermore, one that runs at the old home may not run after this heavy creature is installed in the new home. The old Servel is also a gas hog. The Sibir model, or a Dometic-Servel, uses 5 gallons of propane per month in a seventy degree environment. Though it uses a little more propane in the hottest climates, it will still cool and freeze in the summer. A Sibir can pay for its replacement of an inefficient old Servel in a few years with its lower gas use.

Photo 6-1 Servel™ Propane Refrigerator

PV/GEN HYBRID

A natural hybrid of a PV system is a PV/GEN hybrid. The weakness of a small PV system is its dependence on a balance of the frequency of sunshine to the frequency of use. A special construction project or a winter period of no sun can upset the PV "cash flow". Also a PV system which matches one pocketbook can supply enough electricity for ten months but fall short for two months. The weaknesses of a gasoline generator are noise and inefficiency when running small loads for a long period of time.

In a PV/GEN hybrid, the PV array performs the majority of the charging of the battery bank. Loads are met by the battery bank as needed. When the PV array is delinquent in the charging of the battery bank, the generator is run as a back-up system powering a 12V or 24V large wattage charger. In a short amount of generator running time, this recharges the batteries. The generator, rather than only powering a back-up battery charger, can also power large AC loads directly. In an ideal situation, the generator powers a washing machine and water pump directly, while it also

*Servel™ is a trademark of the Dometic Corporation, Elkhart, IN 46515

powers a battery charger to recharge a low battery bank. The system is being used more efficiently when the generator directly powers the washing machine and charges the batteries, rather than when it first charges the batteries and then powers the washing machine with an inverter. In the latter case, extra electricity is consumed by the inefficiency of the battery charger, the battery bank, and the inverter.

A generator runs most efficiently when the loads approach the generator's rated output. Therefore, it is better to run a washing machine and battery charger together, rather than separately for a longer total time. A true PV/GEN hybrid is more than a PV system with a back-up charger. The AC fuse box is wired to a manual or automatic transfer switch which receives power from either the inverter or the generator. A battery charger is powered directly from the generator before the transfer switch. It is permanently attached to the battery bank. In the normal situation, the transfer switch connects power from the inverter to the house AC fuse box. When the generator is running, it constantly charges the batteries, and the transfer switch disconnects the inverter and connects the AC power from the generator to all AC house circuits. Therefore, whenever the generator is on to charge the batteries it runs all AC loads. Conversely, whenever the generator is running an AC load it automatically charges the batteries. A battery charger in this application must automatically taper the charging rate to prevent overcharging the batteries. (See Figure 6-2.)

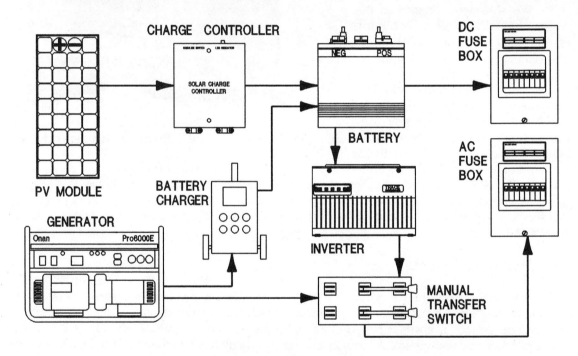

Figure 6-2 PV/GEN Hybrid Using Separate Battery Charger

Heart Interface popularized an inverter model which doubled as a battery charger. In the PV system, the inverter changes the low voltage DC to 120V AC. When the generator is started, the inverter senses the new source of current and automatically disconnects the inverter from the AC fuse box and connects the AC fuse box instead to the generator. Now all household AC loads are powered directly by the generator. The inverter effectively runs in reverse, changing the 120V AC to low voltage DC to charge the battery. When the generator is shut down, the inverter resumes its primary role. (See Figure 6-3.)

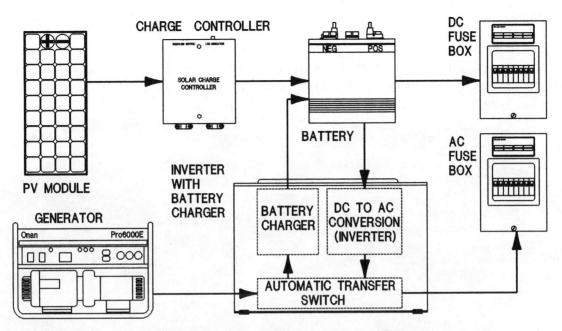

Figure 6-3 PV/GEN Hybrid Using Inverter with Built-in Battery Charger

The Trace inverter also has an inverter/charger model. We prefer this unit to the original Heart unit because it has a programmable battery charger. The voltage taper point and the amperage charging rate are user selectable by setting DIP switches. This means that on a larger battery bank the taper point can be set high, thus always putting out the full 100 amps on the 12V model and the full 52 amps on the 24V model. The Heart model does a fairly quick taper and the amperage output is then only about half of the rated maximum. This of course means the generator needs to run twice as long.

Any battery charger is dependent on the peak voltage of the AC source. A small generator cannot maintain a good high peak voltage and thus puts out less than the rated output for the charger. A 50 amp 24V charger will put out its maximum on a good quality (thus an accurately rated) 3500 watt generator. Of course, if the generator is already significantly loaded, it will cut down on what it can give to the charger.

Frequently we receive a call from a customer who is having a problem with his PV/GEN hybrid. The generator will not run both a large load and the battery charger even though both of them should be well less than the 3500 watt output of the generator. This is the problem. The generator supplies 1750 watts out of one receptacle and the other 1750 watts out of the other receptacle. Usually this generator will have a larger, more heavy duty 30 amp receptacle on the control panel that will supply the full 3500 watts at 120 volts. There will be a switch labeled FULL POWER 120 VOLTS in one position and 240 VOLTS in the other position. It is in the 240 VOLT position that you will receive half power at each receptacle and 240 volts from the 240 volt receptacle. You will need to position this switch for full power 120 volts and use the full power 120 volt receptacle.

Some generators will need to be rewired on an internal terminal strip to accomplish full 120 volt output. A few cheaper generators will never supply full 120 power. These generators should be avoided. Please consult your generator manual or your generator dealer for specific wiring information.

Generators' and inverters' specifications vary depending on the model and the manufacturer. They output potentially deadly voltage. Installation and design of this equipment and their systems should be carried out by qualified personnel. This chapter is not an installation guide.

Photo 6-2 Double Trace Inverter/Battery Charger

Photo 6-3 PV Home (Courtesy of Backwoods Solar Electric Systems)

Chapter 7

Buying a PV System

A system may be purchased from one dealer, who will analyze the site, provide all the components, and install the system. This is the easiest, quickest, and most expensive way to get a PV system. Or, you can research the necessary material, design your own system, and search out the best price on each component. Neither method works ideally for the remote site home owner. The first is too expensive and provides a system that is a mystery. It is best to take part in the design and installation so that you can be a knowledgeable service person for your own system.

The second approach requires a lot of time and energy to become an expert. It is also too expensive for all but the cleverest. You may get the best price, but you also buy your own mistakes. If you buy the wrong inverter, you live with your mistake or pay for it in a future trade-in. It is best to buy the majority of the components from a local dealer who will stand by the equipment, call the manufacturer and facilitate a speedy warranty repair, or research the question you have about a battery rating.

In this specialty business, a local dealer will almost never be the next town over. What is important is a supplier who knows your area and who can ship you something in a day or two. In the worst case, you can take a long ride to his shop.

Certainly shop for price. Some dealers sell components at list price and others sell at a discount. As a system increases in size, the price of the individual components decreases. We recommend that you seek a fair price and bargain for some free assistance. Whether it is included in the price of the components or paid for additionally, you will need an analysis of your site, an evaluation of your system size and load projections, support advice during your installation, and a system check before you throw the final switch. Though this sounds ominous, most of it can be done in telephone consultations.

The next subject is always a little hard for some customers to grasp. The best PV design is the best deal going. Over the years, we have had hundreds of customers buy equipment to add onto existing systems. We have almost never seen a system that did not need improvements in basic design. Often the system was homeowner designed, based on poor information. Sometimes the information came from a poor dealer. Sometimes it was information that had since grown old.

A dealer who has a lot of experience has more knowledge to go on. More and more we hear of dealers who are in business and when questioned have only the experience of their own home and a few systems. They are trying to break into the business with no real experience or background. They promise to come over and help you install your system. Unfortunately, they have little help to offer.

Most dealers have the same products for sale. Perfection is a system that has the best balance of components. The system with a proper balance of components for your region may outperform a stock system for the same price.

Another asset of a dealer is his ability to troubleshoot problems. We have found that almost never is there a problem with equipment. However you, the buyer, still need help finding what setting is wrong or where you or your electrician has erred. This can be supplied by telephone or by mail with information bulletins.

PV dealers are not listed on Wall Street. The small ones normally have a second profession and a demand on their time. Larger ones offer cheap mail order prices and often a lot of free advice. They offer more information for free than your auto repairman or your Sears repairman. It is not particularly ethical to use their time, telephone bill, and expertise to then save $75 on a price from a mail order company that did not supply the same services. In the future, if something malfunctions or a question needs to be answered, you may need assistance.

Some states still offer tax credits for a PV system. Tax credits may be taken on all PV components including a back-up generator and battery charger. The system must be installed at the buyer's principal residence. In our home state of Massachusetts a 35% tax credit can be applied to state income tax. Also in Massachusetts, this same equipment is sales tax exempt.

Do not confuse a tax credit with a tax deduction. A tax credit reduces the tax you owe at the end of your income tax statement. It also creates a rebate if the tax you owe has already been withheld. If you purchase $1000 of PV equipment you will receive $350 of Massachusetts tax credits. If you owe $350 in taxes you will owe zero dollars after you apply your tax credits. The maximum total tax credit in Massachusetts is $1000.

Chapter 8

Siting the PV System

Roofs are natural places to put PV modules. The height factor decreases the chances that a tree line or ridge will shadow the array. A non-adjustable rack on a roof, with an angle compatible to its latitude, requires only a little additional hardware besides the frame itself. Adjustable frames can also be installed on a roof if the owner can conveniently reach them three times a year to adjust the angle.

Wall mounts offer a simplicity of design if the wall faces south. Again the house becomes the very sturdy, already existing, support structure. Modules are easier to reach for seasonal adjustment. Normally it is a shorter distance from the array to the battery bank, thus minimizing wire length, size, and cost. A wall mounted array is more sheltered from lightning and high wind. One must note the overhang of the eaves and consider the sun angle in the summer months to be sure summer shading will not result in the necessity of moving the array.

Photo 8-1 Nestico-King Home (Courtesy of Fowler Solar Electric Inc.)

Photo 8-2 Derby-Kilfoyle Home (Courtesy of Fowler Solar Electric Inc.)

A ground mount is the easiest structure to build yourself. A few large pipes or pressure treated posts anchored in the ground will support the frame. Adjustment or snow removal is convenient. The array can be placed anywhere in the yard and faced easily in any direction. The distance to the battery bank is, of course, a primary concern.

For any mounting design the array should face due south. This is not the south of a compass. Any town, county, or geologic survey map of the area will indicate magnetic north and the number of degrees deviation from true north. (See Appendix.) True south will always be the direction toward the sun when it is highest in the sky between noon and 1 PM. Optimum insolation occurs if the array receives full sun from 9 AM to 3 PM. An exposure of 8 AM to 2 PM or 10 AM to 4 PM is virtually as good. A site may have an easterly hill or as in the following example a westerly hill. When the array is pointed due south the westerly hill blocks afternoon sun while the easterly valley gives more morning sun. In this case the array should be directed up to 25 degrees eastward to maximize the daily insolation. (See Figure 8-1.) An array that is facing 25 degrees east of south or 25 degrees west of south will receive 90% of the sun available to a perfect southern exposure.

In the summer at 42 degrees latitude, the sun rises slightly north of due east, is high over head at noon, and sets slightly north of due west. In the winter the sun rises in the southwest, arcs not very high in the sky, and sets in the southwest. (See Figure 8-2.) At 42 degrees latitude, the summer noonday sun is 71.5 degrees above the horizon. The winter sun is 24.5 degrees above the horizon. This is a 47 degree difference.

As one moves south from the 42 degrees latitude in the example above, the winter noonday sun is higher in the sky and the summer noonday sun is always the same 47 degrees higher. By the time one gets all the way to the equator, the summer sun is actually 23.5 degrees north of the overhead position. Or, more precisely, it is 123.5 degrees above the horizon. In the appendix of this book you will find a table for seasonal positions of the sun in relation to latitude.

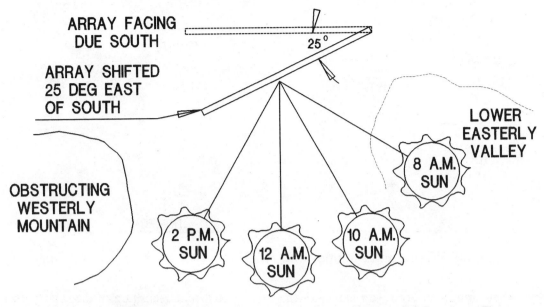

Figure 8-1 Six Hours Of Exposure of an Array Shifted 25 Degrees East

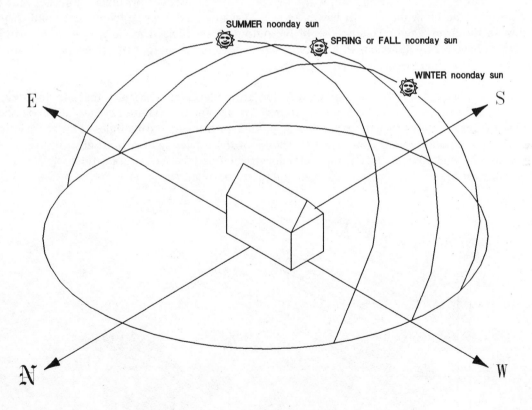

Figure 8-2 Seasonal Path of the Sun at 42 degrees Latitude

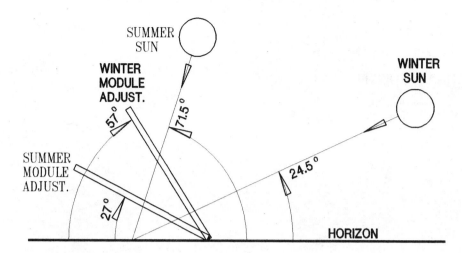

Figure 8-3 Winter and Summer Sun Angles and Array Angles at 42 Deg. Lat.

When siting your array, you must plot the summer and winter position of the sun's path to assure year round exposure. Potential winter tree shading may be decreased by the absence of deciduous leaves, but this shading still will be greater than no trees. How much more depends on the density of the branching. If you have a tree or structure that is close to the array and casts a shadow on the array, then virtually 50% of the output of the shaded module or modules will be lost! If you have a 24 volt system with pairs in series, 50% of the output of the whole pair will be lost with a shadow on any part of either module.

Snow reflection should be considered in the array placement. Modules that are adjusted to the low winter sun of northern regions can receive up to 25% additional insolation by snow reflection from an unobstructed flat stretch of ground in front of the modules. This is easiest to visualize for a ground mount with a ten to twenty foot flat reflecting area in front of it; but it also applies to wall and roof mounts if the flat unobstructed snow area is proportionately longer to compensate for the proportionately higher mounting of the array. (See Figure 8-4.)

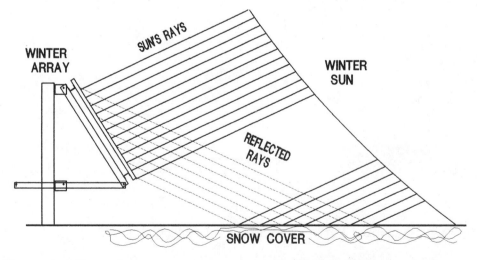

Figure 8-4 Extra Sunlight on Array from Snow Reflection

Photo 8-3 Cook Home (Courtesy of Fowler Solar Electric Inc.)

Winter in colder climates means snow. Modules will clean themselves off naturally if they have an angle around 45 degrees; and faster if they have a 60 degree angle. The snow slumps and exposes some of the module to the sun. This insolation warms the dark surface of the module and the snow slides off. Nothing is, of course, perfect. The problem usually comes when an ice storm turns to a snow storm. The snow doesn't slump off the ice and the snow reflects the sun and does not get absorbed by the dark surface underneath. The solution is to brush off the snow and then the sunlight passes through the ice and warms the modules. If the modules are accessible and the battery bank is down, hand sweep the snow off as soon as possible to harvest every watt possible.

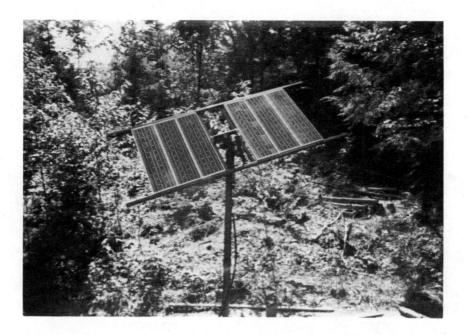

Photo 8-4 Tracker (Courtesy of Fowler Solar Electric Inc.)

Chapter 9

Sizing a PV System

A PV system must be large enough to produce enough watt-hrs to power a load. In the simplest situation, one module that produces 50 watts will power one 50 watt light bulb for every hour that the sun shines. Most light bulbs are not used in the daytime but in the evening. Furthermore, one light bulb is not matched to one module. Instead, a number of modules charge a battery bank all day long, and then this battery bank powers loads in the evening. If an array produces 1000 watt-hrs, only 1000 watt-hrs are available to power the load. The specific load can be 1000 watts for one hour or 100 watts for ten hours. A battery bank is really a bank, deposits equal withdrawals. The PV array steadily charges the batteries during the daylight while the load is withdrawn anytime whether in large wattages over a short time, or in small wattages over a long time.

The first step in designing a PV system is to determine the daily usage in watt-hrs. Next, an array is sized large enough to produce the same watt-hrs as the daily load, and a battery bank is sized large enough to hold the daytime array output and return it at night to power the loads. For example, a 100 watt-hr load is matched to a battery bank which stores 100 watt-hrs, which is matched to a PV array which produces 100 watt-hrs. Unfortunately, this example only works under the ideal condition of 100% efficiency. In reality, the modules lose efficiency as they increase in temperature, the batteries are only 80% efficient in their conversion of electrical energy to chemical energy to electrical energy, and an inverter has a 15% loss of energy for any AC load.

In summer, days are long and nights are short. This reverses in winter. When days are long, the PV array receives more light and produces more electricity. When days are short, less PV electricity is produced, and the longer nights result in greater loads due to increased lighting demands. Clear weather is less common in winter, further reducing the average insolation available.

In the appendix of this book, you will find an appliance list to aid you in the wattage requirements of the most common appliances to be used in a remote site PV home. Also in the appendix, you will find a blank worksheet for sizing your PV system. The first section of this worksheet helps you estimate your daily loads. Photocopy the blank worksheet for your own personal worksheets. In the following pages of this chapter, you will find the above mentioned PV system worksheet filled out with the hypothetical system we will design.

To accurately size a PV system for a remote site home, it is necessary first to determine the loads. Your loads will change throughout the year. If you will have no backup charging system, and all loads are essential, you should fill out the load section of the worksheet for the worst case month of winter when the least average sunlight per day is available and the greatest loads are needed.

If you have a backup charging system, you may fill out the worksheet with load estimates for the average use of the year. During the worst part of the year, additional loads can be supplied by the backup. It is often helpful to fill out the worksheet for several scenarios. After pricing a PV system for each case, you will be able to make a well informed decision.

For the following sample load estimates, we have determined a daily average load. In the AC section we have increased the load by a total of 15% for inverter losses. This load worksheet has no compensation for battery inefficiencies. These will be accounted for in our method of determining usable output of a PV module in the system sizing.

A module produces its peak wattage under full insolation of noontime on a perfectly clear day. In the morning or afternoon hours, the sun is not as intense and the angle of the sun to the module is poor; so the module may only produce half of its peak wattage. Insolation data for the PV industry is presented in peak sun hours. These can most easily be thought of as noontime full sun hours. For a period of ten hours of clear day sunlight, you may only receive six peak sun hours.

In the appendix you will find four maps of the United States. These maps will give you the average daily peak sun hours for your location. If you have five average daily peak sun hours, over a season, then each day will average five noontime full sun hours. The amount of insolation of the five peak sun hours may be spread over ten hours, or the sun may not shine for two days but then shine for the next six days restoring the average. Each map is labeled with the season it represents and the array angle adjustment for that season.

As we work through the following processes of sizing a PV system, we will use the site of Fowler Solar Electric Inc. for a location. The sample PV module will be a 50 peak watt medium voltage module.

FSE SAMPLE SYSTEM

LOCATION: Worthington, MA
LATITUDE: 42 degrees
MODULE: 50 peak watts, medium voltage

To find the seasonal array angle adjustments for an adjustable frame we use the following four formulas.

SUMMER ARRAY ANGLE = Your lattitude minus 15 degrees
FALL ARRAY ANGLE = Your latitude
WINTER ARRAY ANGLE = Your latitude plus 15 degrees
SPRING ARRAY ANGLE = Your latitude
FIXED ARRAY (non-adjustable) = Latitude plus 15 degrees

ADJUSTABLE ARRAY FRAME

SUMMER ARRAY ANGLE = Latitude minus 15 degrees
SUMMER ARRAY ANGLE = 42 - 15 = 27 degrees

FALL ARRAY ANGLE = Latitude
FALL ARRAY ANGLE = 42 degrees

WINTER ARRAY ANGLE = Latitude plus 15 degrees
WINTER ARRAY ANGLE = 42 + 15 = 57 degrees

SPRING ARRAY ANGLE = Latitude
SPRING ARRAY ANGLE = 42 degrees

FIXED ARRAY = Latitude + 15 degrees
FIXED ARRAY = 42 + 15 = 57 degrees

AVERAGE DAILY MODULE OUTPUT

Our 50 watt module will theoretically produce 50 watts. Unfortunately, there are battery losses when this energy is stored in the battery bank to later power a load. To simplify our PV sizing calculations, we will derate the modules initially to find what usable electrical energy they will have to power a load on the load side of the batteries. This will, in effect, eliminate the need for compensating for battery inefficiencies in the load estimate of the worksheet. To calculate the usable output for a module we must first derate its rated output to get the output for normal operating voltage of the battery bank. The sample module is rated at 50 watts at typical load, or 3.13 amps at 16 volts. Unfortunately, you actually store the 3.13 amps in the battery which supplies your loads at an average 12 volts.

WATTS = AMPS X VOLTS
WATTS = 3.13 X 12 = 37.56

The 50 watt module will really only supply 37.56 watts that you can take from the battery bank. A higher voltage module is rated at 3.17 amps at 17.4 volts at typical load. This is 55 watts. However the usable wattage from the batteries is 3.17 amps at 12 volts. This is remarkably similar to the previous medium voltage module.

WATTS = 3.17 amps X 12 volts = 38.04

The higher voltage module will not give you more wattage that you can use. However, in hot climates the medium voltage module will not perform at a high enough voltage as the temperature brings its operating voltage down. This is where the higher voltage module is meant to be used. Of course the temperature brings its normal operating voltage down to around 16 volts like the medium voltage module in a temperate climate.

USABLE MODULE OUTPUT = AMPS (at typical load) X NOMINAL MODULE VOLTAGE
USABLE MODULE OUTPUT = 3.13 amps X 12 volts = 37.56

To find the average daily module output, you will multiply the usable module output by the number of hours of sun. This will be the number of average peak sun hours. We find this on the maps in the appendix. The product of the above will be the average daily output in watt-hrs.

AVERAGE DAILY USABLE MODULE OUTPUT = USABLE MODULE OUTPUT X AVERAGE DAILY PEAK SUN HOURS

If you will use an adjustable array you may calculate the average daily peak sun hours for each season and each different adjustment.

If you will use a fixed array for a year round home we recommend an angle of your latitude plus 15 degrees. This will normally give the best possible output for the worst season of the year, winter. Spring and fall will have a only a slightly smaller output than they would at their adjustable array adjustment of latitude. Summer output will be less than the latitude minus 15 degree adjustment, but your system should be full all the time anyway as the loads should be least at this time of year. The yearly average output of the non-adjustable array angle (latitude plus 15 degrees) should be close to the average of the fall and spring output figures for an array angle of your latitude.

In the appendix there is data of the average daily peak sun hours for most major cities in the United States. This data is given for each month and for the whole year for angle adjustments of latitude plus 15 degrees, and latitude minus 15 degrees. You may use this data instead of the data from the four maps whenever you feel it will give you a more accurate analysis of your specific design and location. You will notice that the data from these tables is more conservative than the data from the maps. The data from the tables is for major cities. We surmise that this data reflects the poorer insolation of the polluted city skies.

FSE SAMPLE SYSTEM

AVERAGE DAILY MODULE OUTPUT

To calculate the average daily module output, first turn to the appendix and find the four maps. Find the average daily peak sun hours for each of the four seasons for your location and list them.

ADJUSTABLE ARRAY

SUMMER AVERAGE DAILY PEAK SUN HOURS = 5.75
FALL AVERAGE DAILY PEAK SUN HOURS = 4.5
WINTER AVERAGE DAILY PEAK SUN HOURS = 2.5
SPRING AVERAGE DAILY PEAK SUN HOURS = 4.3
FIXED YEARLY AVERAGE DAILY PEAK SUN HOURS = SPRING VALUE [4.3] + FALL VALUE [4.5] / 2 = 4.4

AVERAGE DAILY USABLE MODULE OUTPUT = USABLE MODULE OUTPUT X DAILY AVERAGE PEAK SUN HOURS

SUMMER AVER. DAILY USABLE MOD. OUTPUT = 37.56W X 5.75HRS
SUMMER AVER. DAILY USABLE MOD. OUTPUT = 216 WATT-HRS

FALL AVER. DAILY USABLE MOD. OUTPUT = 37.56W X 4.5HRS
FALL AVER. DAILY USABLE MOD. OUTPUT = 169 WATT-HRS

WINTER AVER. DAILY USABLE MOD. OUTPUT = 37.56W X 2.5HRS
WINTER AVER. DAILY USABLE MOD. OUTPUT = 94 WATT-HRS

SPRING AVER. DAILY USABLE MOD. OUTPUT = 37.56W X 4.3HRS
SPRING AVER. DAILY USABLE MOD. OUTPUT = 162 WATT-HRS

FIXED YEARLY AVER. DAILY USABLE MOD. OUTPUT = 37.56W X 4.4HRS
FIXED YEARLY AVER. DAILY USABLE MOD. OUTPUT = 165 WATT-HRS

SYSTEM ARRAY SIZING

We now have a load estimate from the sample worksheet that follows and the average daily usable module output data for Worthington, MA from above. We simply need to divide the average daily load by the average daily usable module output to determine the number of modules needed to power our loads. For brevity, we will size this system for the yearly average daily output of a fixed array at the winter array angle. You may spend more time and work the calculations for all adjustments of an adjustable array.

NUMBER OF MODULES = AVERAGE DAILY LOAD / AVERAGE DAILY USABLE MODULE OUTPUT

NUMBER OF MODULES = 1318 WATT-HRS / 165 WATT-HRS
NUMBER OF MODULES = 8

BATTERY BANK SIZING

We will do battery sizing by a shortcut method based on longtime professional experience and on calculations not shown here. Match one 50 watt module to 1500 watt-hrs of battery storage with the battery storage rated at a 20 hour discharge rate. This is also 30 watt-hrs of battery storage for each watt of module rated output. This will be adequate storage for northern climates that experience long periods of no-sun in winter. For areas that do not have long periods of no-sun, the battery storage can drop to as low as 1000 watt-hrs per 50 watt module or 20 watt-hrs per watt of module rated output.

BATTERY BANK = 1500 WATT-HRS X NUMBER OF 50W MODULES

[This can alternately be expressed as:]

BATTERY BANK = 30 WATT-HRS X RATED MODULE WATTAGE X NUMBER OF MODULES

NUMBER OF BATTERIES = BATTERY BANK WATT-HRS / BATTERY WATT-HRS

Obtain battery ratings for deep cycle lead acid batteries at a 20 hour discharge rate. This will be in amp-hrs. Multiply the amp-hrs by the battery voltage for each battery to find how many watt-hrs each battery holds.

BATTERY WATT-HRS = BATTERY AMP-HRS X BATTERY VOLTAGE

FSE SAMPLE SYSTEM - BATTERIES

BATTERY BANK SIZING

6V BATTERY
200 AMP-HRS AT 20 HR DISCHARGE RATE

BATTERY WATT-HRS = 200 AMP-HRS X 6 VOLTS
BATTERY WATT-HRS = 1200 WATT-HRS

BATTERY BANK = 30 WATT-HRS X MODULE WATTS X NUMBER OF MODULES

BATTERY BANK = 30 WATT-HRS X 50 X 8 MODULES
BATTERY BANK = 12000 WATT-HRS

NUMBER OF BATTERIES = BATTERY BANK WATT-HRS / BATTERY WATT-HRS

NUMBER OF BATTERIES = 12000 WATT-HRS / 1200 WATT-HRS
NUMBER OF BATTERIES = 10 BATTERIES

The above system has been designed to satisfy the electrical requirements of an estimated average daily load. Often you will need to size a system by price and then be able to calculate what you will have available to power loads. The following is an alternate method.

PART A: LOAD ESTIMATES

DC APPLIANCE	HRS/DAY	X	WATTS	=	WATT-HRS/DAY
4-15 W Fluorescents	4	×	60 W	=	240
Tape Player	2	×	10 W	=	20
2 25 W Lamps	2	×	50 W	=	100

DC AVERAGE DAILY LOADS ___360___

AC APPLIANCE	HRS/DAY	X	WATTS	=	WATT-HRS/DAY
1 20" Color TV	3	×	60 W	=	180
Submersible Pump	.3	×	1000 W	=	333
Computer	1	×	80 W	=	80
2 AC 15 W Fluorescents	3	×	30 W	=	90
Blender	.1	×	300 W	=	30
Vacuum Cleaner	.25	×	480 W	=	120

subtotal AC AVERAGE DAILY LOADS ___833___

[15% of above] 15% INVERTER ALLOWANCE ___125___

TOTAL AC AVERAGE DAILY LOADS ___958___

DC AVERAGE DAILY LOADS ___360___

TOTAL [AC AND DC] AVERAGE DAILY LOADS ___1,318___

PART B: PV SYSTEM SIZING WORKSHEET

SECTION 1 - SITE DATA AND MODULE SPECIFICATIONS
LOCATION _Worthington, MA_   LATITUDE ___42°___
MODULE MODEL _med. Volt._ PEAK WATTS ___50___ NOMINAL VOLTAGE _12 V_
AMPS at typical load ___3.13___

SECTION 2 - ARRAY ANGLE ADJUSTMENTS
LATITUDE (-) 15 degrees = SUMMER ARRAY ANGLE A _27°_
LATITUDE   = FALL   ARRAY ANGLE B _42°_
LATITUDE (+) 15 degrees = WINTER ARRAY ANGLE C _57°_
LATITUDE   = SPRING ARRAY ANGLE D _42°_
LATITUDE (+) 15 degrees = FIXED ARRAY ANGLE C _57°_

SECTION 3 - USABLE OUTPUT OF MODULE
MODULE AMPS at typical load E _3.13_ X NOMINAL MODULE VOLTAGE _12_
= USABLE MODULE OUTPUT in watts F ___37.56___

SECTION 4 - AVERAGE DAILY PEAK SUN HOURS from maps in appendix
SUMMER AVERAGE DAILY PEAK SUN HOURS at lat.(-)15 deg. G   5.75
FALL   AVERAGE DAILY PEAK SUN HOURS at lat.   H   4.5
WINTER AVERAGE DAILY PEAK SUN HOURS at lat.(+)15 deg. I   2.5
SPRING AVERAGE DAILY PEAK SUN HOURS at lat.   J   4.3
FIXED   AVERAGE DAILY PEAK SUN HOURS = (FALL A.D.P.S.Hrs.
H 4.5   + SPRING A.D.P.S. Hrs. J 4.3   / 2 = K   4.4

SECTION 5 - AVERAGE DAILY USABLE MODULE OUTPUT
SUMMER A.D.P.S. Hrs. G 5.75 X USABLE MOD. OUTPUT F 37.56
= SUMMER AVER. DAILY USABLE MOD. OUTPUT L 216

FALL   A.D.P.S. Hrs. H 4.5 X USABLE MOD. OUTPUT F 37.56
= FALL   AVER. DAILY USABLE MOD. OUTPUT M 169

WINTER A.D.P.S. Hrs. I 2.5 X USABLE MOD. OUTPUT F 37.56
= WINTER AVER. DAILY USABLE MOD. OUTPUT N 94

SPRING A.D.P.S. Hrs. J 4.3 X USABLE MOD. OUTPUT F 37.56
= SPRING AVER. DAILY USABLE MOD. OUPUT O   162

FIXED   A.D.P.S. Hrs. K 4.4 X USABLE MOD. OUTPUT F 37.56
= FIXED   AVER. DAILY USABLE MOD. OUTPUT P 165

SECTION 6 - ARRAY SIZING
AVERAGE DAILY LOADS 1318 Watt Hrs/ USABLE MODULE OUTPUT Fixed 165
= NUMBER OF MODULES Q   8

SECTION 7 - BATTERY BANK SIZING
BATTERY   VOLTAGE   6 X AMP-HRS at 20 hr. rate 200   = BATTERY
SIZE in watt-hrs R   1200

30 WATT-HRS X MODULE PEAK WATTS 50 X NUMBER OF MODULES Q 8
= BATTERY BANK WATT-HRS S   12,000

BATTERY   BANK S 12,000   / BATTERY SIZE R 1200   = NUMBER OF
BATTERIES T   10

SECTION 7 - ESTIMATED ARRAY USABLE OUTPUT
Here you choose a number of modules and a season to get a budget
for loads.
PROSPECTIVE NUMBER OF MODULES U   8
SEASONAL ARRAY ANGLE from Sect.4   SEASON Lat + 15° ANGLE   57°
AVERAGE DAILY MODULE OUTPUT from Sect.5
NUMBER   OF MODULES U 8   X AVER. DAILY MOD. OUTPUT 165 = AVER.
DAILY ARRAY OUTPUT V 1320 Watt Hrs

AVERAGE   DAILY   ARRAY OUTPUT V 1320 Watt Hrs will be the   average
number   of watt-hrs available to power your loads for the   season
and the angle you have chosen for your prospective   system. You
may repeat the calculations for another season. Use Section 7 for
sizing battery bank for prospective system.

Disregard the sample load worksheet we have filled out. Keep the latitude and frame adjustment data and the average daily module output computations. We will now multiply the average daily module output by the number of modules in your prospective array and get the average daily array output. You will then be able to see if you can size your loads to fit this budget.

PROSPECTIVE SYSTEM:

EIGHT MEDIUM VOLTAGE 50W MODULES
FIXED ARRAY: 57 DEGREES
YEARLY AVERAGE DAILY USABLE MODULE OUTPUT: 165 WATT-HRS

AVERAGE DAILY USABLE ARRAY OUTPUT = AVERAGE DAILY USABLE MODULE OUTPUT X NUMBER OF MODULES

AVERAGE DAILY USABLE ARRAY OUTPUT = 165 WATT-HRS X 8
AVERAGE DAILY USABLE ARRAY OUTPUT = 1320 WATT-HRS

Battery bank sizing is performed exactly as in the system sized from a load estimate, only you will use the prospective number of modules for the number of modules.

In our ten years of researching, living with, designing, and installing remote site home PV systems in New England, we've rarely received an accurate enough estimate of load to size a PV system by any of the accepted mathematical procedures. Remote site home systems are rarely sized precisely by load. An owner has a maximum amount of money to spend and needs to know what a system will produce. This system output is then budgeted until the owner can afford system expansion.

There are many more complicated ways to size systems. When PV systems power constant loads such as a street lamp, they are very accurate. The load is always precisely the same. In a remote site home the load is only an estimate that is poor in the first place and worse when company visits. In New England, the average daily sunlight in winter is only half of what it is in summer. A system accurately sized for winter is producing twice the current needed in summer. For some people, an obvious solution is to decrease the system size (and of course the capital investment) and conserve in winter. Another option is to supplement the low power time of winter with a gasoline generator and a battery charger.

On the next several pages, five different remote site home PV systems are priced and described. These are systems we have owned, monitored, or installed. They all assume seasonal adjustment of the PV array to the sun's angle above the horizon. The estimates of loads for winter and summer apply to the insolation of New England. You can, of course, work out load estimates for these systems using the previous methods.

The module prices are based on $6.60 a watt. Module prices vary with dealers and manufacturers. Module prices also vary with total system size purchased and with quantity of modules purchased. These figures should give you a usable estimate.

SYSTEM NUMBER ONE   12V   100 peak watts

2	50 watt modules	$660
4	6V batteries 200 amp-hr	320
1	30 amp charge controller	89
4	battery interconnects	14
2	module interconnects	4
1	array disconnect circuit breaker	35
1	charge controller disconnect box	35
		Total $1157

OPTIONAL COMPONENTS

600 watt inverter $550 (with 25 amp battery charger add $100)

SUMMER USE

12V TV

12V stereo, tapedeck, radio

6 - 13 watt 12V fluorescent lights 4-6 hrs/day

WINTER USE

4 - 13 watt 12V fluorescent lights 4 hrs/day

TV or stereo - limited use or substituted for a light

OPTIONS

1. AC generator for AC loads and a battery charger to charge battery bank.

2. Greater weekend use balanced by lesser weekday use.

This small system will release a remote site home from the world of gas and oil lamps. It will also provide music, which is often the other motivating reason for seeking an alternative electricity to replace an AC generator. During the long days of summer, a couple will have a surplus of electricity if they use only the low wattage appliances listed. In winter, the system must be budgeted. A backup 12V charger is the answer, preferably one of those fifty pound models with the little wheels. You will need it only once a week and only in winter. This needs to be powered by a small AC generator. If possible, borrow both of them in times of no sun.

A small inverter with high motor surging capabilities (like the Trace 600 watt model) will power a vacuum cleaner, blender, or drill. These are large wattage appliances, so expect to use them only in times of much sun.

This system has larger module to battery ratio than the following systems. This larger battery bank helps to get you through long no sun periods while also providing a good sized bank for supplemental charging.

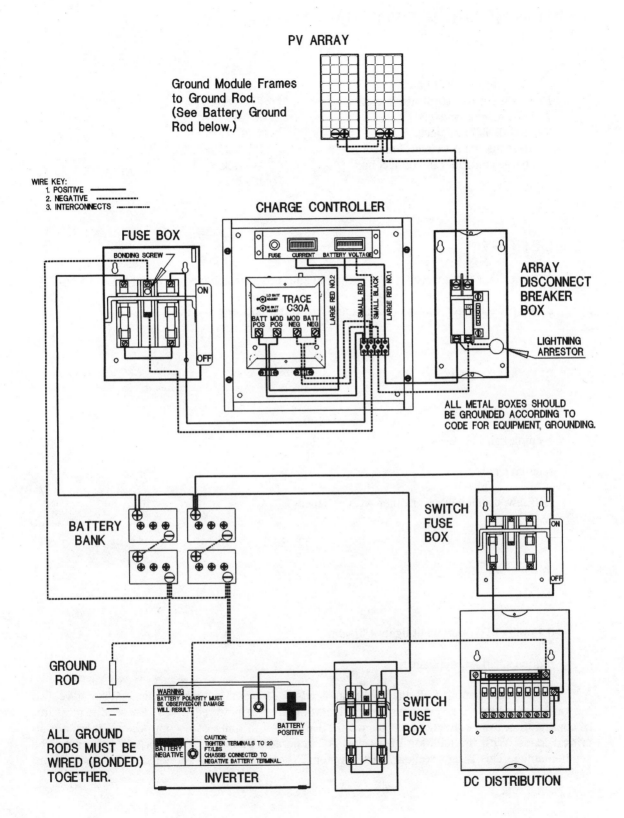

Figure 9-1 System Number One Diagram

SYSTEM NUMBER TWO   12V   200 peak watts

4	50 watt modules	$1320
6	6V batteries 200 amp-hr	480
1	30 amp charge controller/meters	180
7	battery interconnects	24
6	module interconnects	12
1	array disconnect circuit breaker	35
1	charge controller disconnect box	35
		Total $2086

OPTIONS

1. Trace 2000 watt inverter $1090 (with 100 amp programmable battery charger add $220)

2. AC generator for winter AC loads and for winter backup charging.

SUMMER USE

12V TV, stereo, or radio

evening lighting needs

vacuum cleaner

skill saw, drill, blender, or other AC appliances

WINTER USE

12V TV, radio, or stereo

6 13 watt 12V fluorescent lights 4 hrs/day

limited small AC appliances

This system supplies all the electrical needs for several of our customers who have two person households. Water pumping and refrigeration are not powered by PV. Successive measures of conservation can be practiced to balance the PV budget in times of no sun because there are no critical loads. With the addition of a small AC generator and a 12V charger, PV power can be supplemented. This is an excellent starter system. The inverter may be added when finances are available.

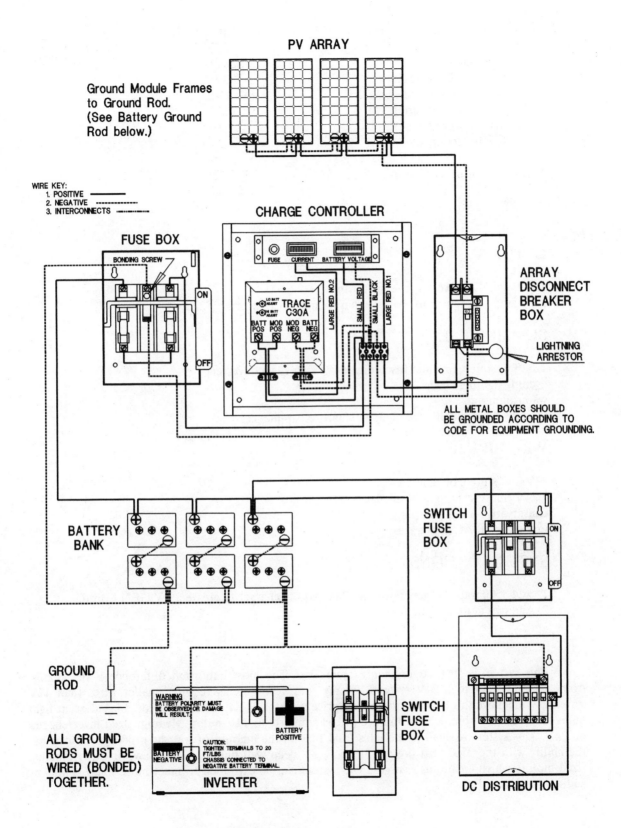

Figure 9-2 System Number Two Diagram

SYSTEM NUMBER THREE   24V   400 peak watts

8	50 watt panels	$2640
12	6V batteries 200 amp-hr	960
1	30 amp charge controller/meters	180
1	Trace 2500 watt inverter	1350
13	battery interconnects	45
10	module interconnects	20
1	array disconnect circuit breaker	35
1	charge controller disconnect box	35
		Total $5265

OPTIONS

1. 100 amp DC inverter fusebox for inverter $100
2. 52 amp programmable battery charger for Trace inverter $220

SUMMER USE

12V TV, stereo, or radio [use efficient 24V to 12V converter]
evening lighting needs
water pumping
vacuum cleaner, skill saw, blender, small AC appliances

WINTER USE

12V TV, stereo, or radio
4 24V fluorescent lights 4 hrs/day
conservative water pumping
limited small AC appliances

SYSTEM OPTIONS

1. Add 3500 watt AC generator and 24V battery charger to create a PV/GEN hybrid.
2. Add two more modules.

This system is 50% larger than System Two. The most important difference is the change from a 12V system to a 24V system. Twenty four volt systems require smaller wires than 12V systems. We believe this is worth the higher price paid for 24V fluorescent lights. Incandescent light bulbs are available from dealers and are priced similarly to 12V bulbs. Medium sized inverters run on 24V. This system is small in modules (and thus in the amount of watt-hr produced), but it has the ability with the 2500 watt inverter, to power a deep well pump, a table saw, or other medium sized motors which create high surges. The system can be expanded to meet more loads simply by adding more modules.

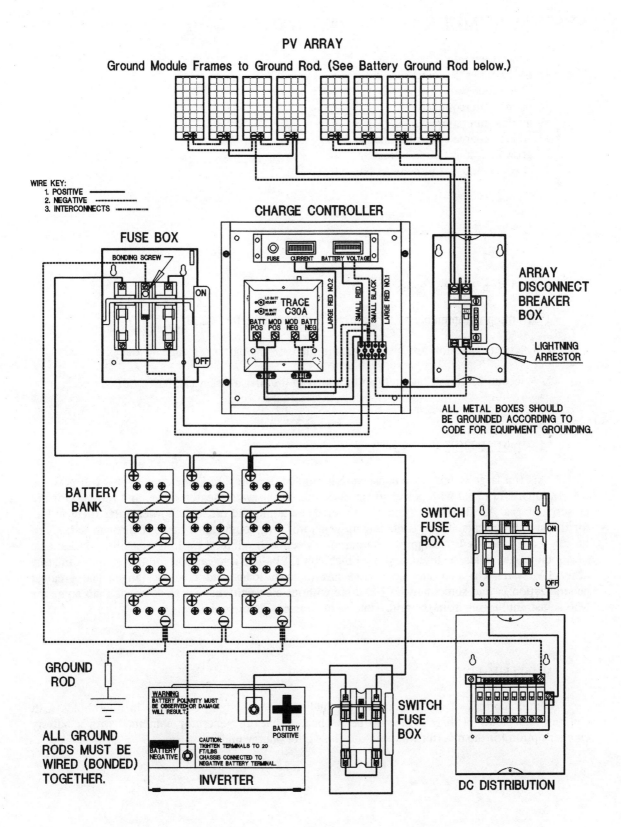

Figure 9-3 System Number Three Diagram

SYSTEM NUMBER FOUR   24V   800 peak watts

16 50 watt modules	$5280
16 6V batteries 200 amp-hr	1280
1 Trace 2500 watt inverter	1350
18 battery interconnects	63
19 module interconnects	38
1 30 amp charge controller/meters	180
1 array disconnect circuit breaker	35
1 charge controller disconnect box	35
	Total $8261

OPTIONS

1. 52 amp programmable battery charger for Trace inverter $220

2. The next expansion for this system is to increase the battery storage by four batteries.

3. A second 2500 watt inverter can be added to the first to create a 5000 watt inverter.

SUMMER USE

This is a large system for a conservation minded, remote site home owner. This will power a conventional household with power to spare as long as the water heater, stove, and refrigerator are powered by gas. An identical system has powered the home and office at Fowler Solar Electric Inc. for the last three years. There is no backup generator. We comfortably make it through periods of no sun. Loads include two large PC computers, a deep well pump, a washing machine, a radial arm saw, a table saw, lights, a stereo, and a 20 inch color TV. The one caution for this system (and any other PV system) is that any household has its own loads and comfort range. The greatest generalization is that someone who has done without electricity has fewer demands than someone who is just unplugging from the grid. Time is, of course, a great equalizer.

WINTER USE

Winter periods of no sun can tax the budget in any system, especially in households with children and plenty of dirty clothes to be washed. A backup generator combined with a battery charger option for the inverter will even out the winter energy budget.

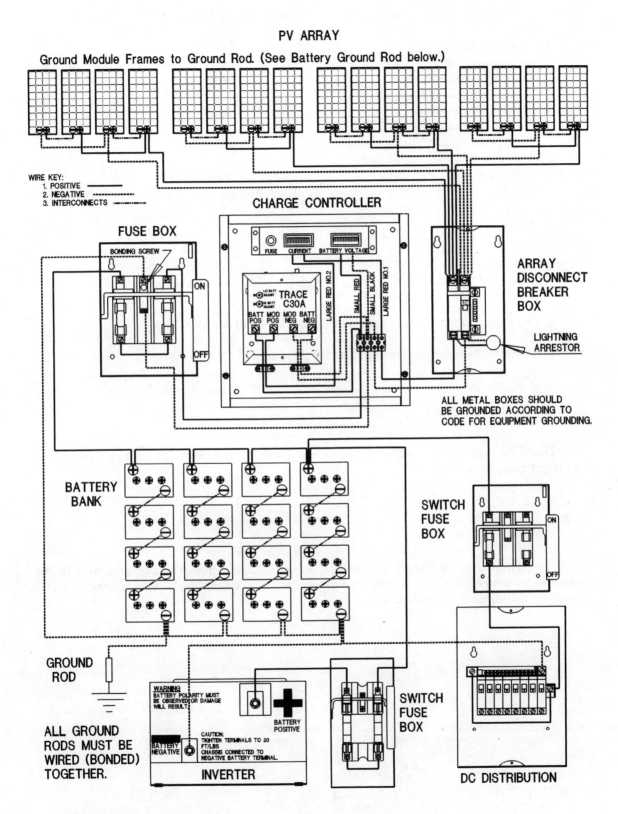

Figure 9-4 System Number Four Diagram

SYSTEM NUMBER FIVE   24V   1200 peak watts

24	50 watt modules	$7920
24	6V batteries 200 amp-hr	1920
2	Trace 2500 watt inverters	2700
1	Trace inverter stacking option	200
28	battery interconnects	98
28	module interconnects	56
2	30 amp charge controllers/meters	360
2	array disconnect circuit breaker	70
2	charge controller disconnect box	70

Total $13,394

OPTIONS

1. Two 52 amp programmable battery chargers for Trace inverters $440

2. Additional batteries may be added for northern climates.

SUMMER USE

This is an expanded version of the previous system. The inverter size has been doubled to meet larger heavy loads.

WINTER USE

This system may have been chosen to eliminate the need for a backup generator in the winter. If the winter loads have increased over the previous system loads, then a generator and inverter charger option is still desirable.

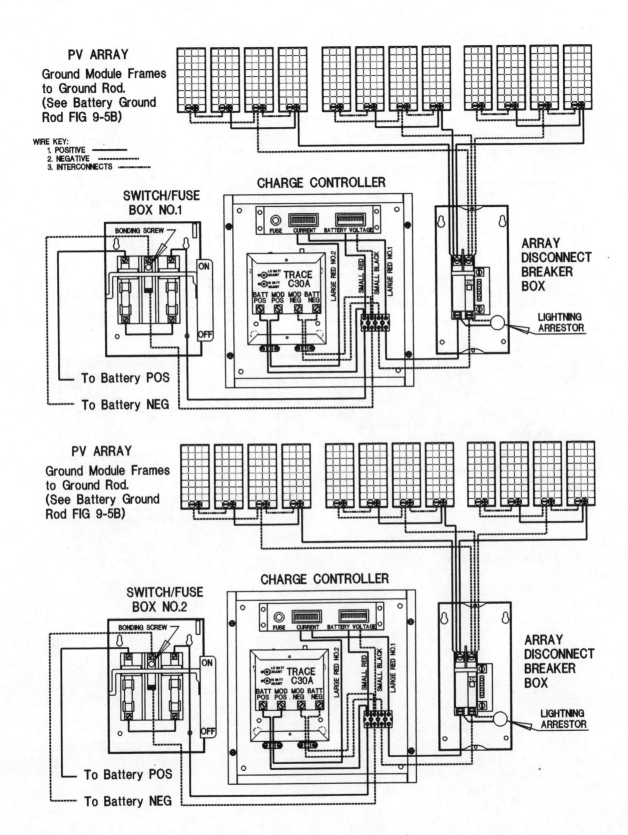

Figure 9-5A System Number Five Diagram Part A.

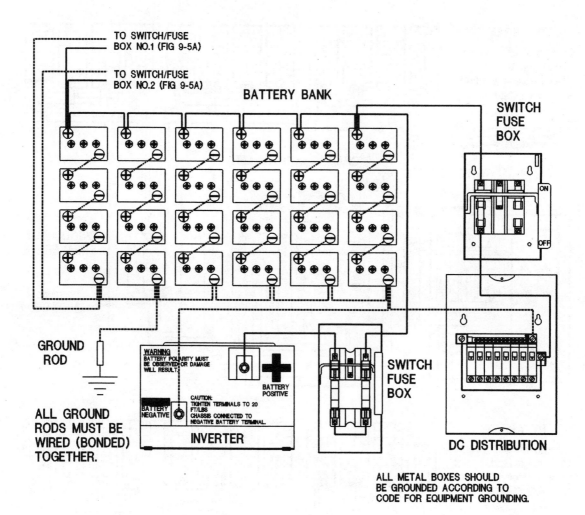

Figure 9-5B System Number Five Part B.

Chapter 10

PV Home Lighting and Wiring

Any new remote site home, or any new construction on a remote site home, should be wired to the specifications of the state electrical code before the walls are closed in. Retrofit wiring is difficult and three times as expensive. Someday you will need the AC wiring; if not you, then the next owner will. A code wired house means no dangerous extension cords. A generator can be hooked to the main service panel to power the whole house. A later expanded PV system can incorporate an inverter.

An AC home following the code is supplied by a 60 amp or larger service panel. This may be a new circuit breaker type or a secondhand fuse type. There will be a main power fuse or breaker and an individual fuse or breaker for each individual circuit. A circuit using 14 gauge wire must be fused at 15 amps and use wall boxes that are two and one half inches deep. A circuit using 12 gauge wire uses 20 amp fusing and requires three and a half inch deep wall boxes. All circuits, receptacles, fixtures, and electrical boxes are grounded back to the service panel where the white or neutral wire is also grounded. The service panel is, in turn, grounded to an eight foot copper clad rod driven into the earth. Grounding is imperative to electrical safety.

Since AC wiring is at 120 volts, long circuit lengths of 12 gauge or 14 gauge wire up to 75 feet result in minimal voltage drop. Normal circuits will power up to 10 receptacles or fixtures. Living areas must have receptacles spaced closely enough together that a six foot cord of an appliance placed anywhere reaches a receptacle. This means one for every eight to ten feet of wall. Hallways and high traffic areas are excluded. Each room must have a switched light fixture or a switched wall receptacle into which a light may be plugged. Long switch legs and additional remote switches are compatible with the low line losses of 120 volts. The circuits to receptacles in bathrooms, dining rooms, kitchens, basements, and those to outside receptacles are always 12 gauge wire with 20 amp fusing. Each circuit will feed only a few boxes. This is for a good reason. These are the receptacles that will power high wattage appliances such as coffee pots, toasters, and power tools. Major appliances such as water pumps are switched and on their own individual 20 amp circuits.

A small PV system for a remote site home will have all or part of its wiring as DC. Like the AC wiring it must have a main fuse or circuit breaker and a fuse or circuit breaker for each individual circuit. All circuitry must be grounded just as in an AC system. Do not be fooled by the apparent safety of low voltage DC. A short will start a fire no matter what the voltage. Have you shorted a wrench recently across your car battery terminals? Furthermore, those efficient DC fluorescents you will invert the low voltage DC to high voltage AC. An ungrounded 12V DC fluorescent fixture is as dangerous as 120 volt wiring.

Any DC circuit must use 12 gauge or larger wire. The large line losses of low voltage dictate a low amperage circuit of a short length of wire. The service panel should be centrally located to facilitate the shortest most direct routes. Each circuit is custom designed using line loss tables. The length of circuit, number of receptacles, and prospective amp draw of appliances are all considered. As a rule of thumb, a 12 volt circuit that is short and direct might have two to four receptacle boxes powering four 25 watt lights. A longer upstairs circuit would be capable of only half the load.

A 25 watt 12V bulb draws 1.2 amps, a 25 watt 24V bulb draws 1.05 amps, and a 25 watt 120V bulb draws .21 amps. (See Chapter 16 for formulas.) Thus, if we fused a circuit at 10 amps we could use four 12V bulbs at 25 watts, nine 24V bulbs at 25 watts, or 48 120V bulbs at 25 watts. But the maximum amperage is only one design determinant; voltage is the other. A 120V line can be ten

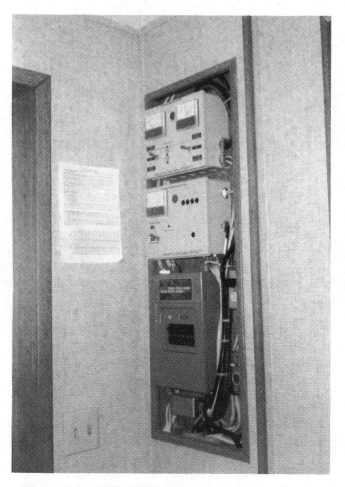

Photo 10-1 DC Distribution (Courtesy of Backwoods Solar Electric Systems)

times as long as a 12V line that draws the same amperage and suffers the same loss. The higher voltage causes the same amperage to flow more efficiently. If you analyze the line loss tables in the appendix you will see that as you increase the amperage, line loss increases. As you decrease the voltage, line loss also increases. If you take a 24V 25 watt bulb and then compare it with a 12V 25 watt bulb, you both decrease the voltage and increase the amperage, making the line loss exponentially worse. Comparing three 25 watt bulbs, one 12V, one 24V, and one 120V, you find that for the same 5% line loss and 12 gauge wire, the 24V bulb can have a wire four times as long as the 12V, and the 120V can have a wire 100 times the length of the 12V.

A DC fuse box will normally only power lights in the remote site home. It can be smaller, simpler, and contain less circuits than its AC counterpart. A main fuse, consisting of a 30 amp auto fuse, followed by circuits each fused by a 10 amp auto fuse will clone the fuse system in your automobile or RV. Some charge controllers as an option contain a 30 amp fuse or circuit breaker for loads, and no additional main fusing is necessary. Alternative energy suppliers usually sell a DC circuit breaker box. A conventional AC fuse box with screw-in glass fuses also works. Any of these options when used in a grounded system, must have a bus bar to ground all negative white wires and all circuit bare wires.

Some strictly low voltage DC PV systems will use a DC refrigerator, water pump, or other high amperage appliance. These appliances should each have their own circuits with proportionately large wiring and fusing.

All switches and circuit breakers in a DC system must be rated for DC. DC wall switches will make a loud click. Silent switches that are most common now, are AC only. As switch contacts are separated, current arcs through the air from one contact to the other. High quality contacts and spring mechanisms are employed in DC switches. AC switches can be lighter duty because the current changes direction sixty times per second reducing arcing. DC current only flows in one direction. Thus one wire is positive always and one wire is negative always. In household systems the black will be positive and the white will be negative. But, beware; in auto or RV systems red is positive and black is negative. Only incandescent lights or heating elements need pay no attention to positive and negative wiring. All other DC appliances, including fluorescent lights, must use polarized wiring.

The *National Electrical Code®* specifies that DC connectors must have a different configuration and be of the locking type. It is not clear to us if this means main connectors only or also includes household DC receptacles. It is pretty safe to assume your inspector will be happy with a different plug configuration. Automotive cigarette lighter plugs are expensive and of poor quality. We recommend light duty 240 volt plugs and receptacles. They are similar to a conventional three prong 120 volt plug except that the two flat prongs of a 240 volt plug are in the same plane rather than parallel to each other. The receptacles are available in duplex forms. Wire all receptacles consistently for positive and negative and you will have a polarized system. By convention, the black wire goes to the copper colored screw and the white wire goes to the chrome screw. When using this three prong plug, contrary to using a two prong plug, the plug cannot be turned upside down causing a reverse in polarity. And most importantly, you will not be able to accidentally plug a DC appliance into an AC receptacle. Remember to wire all fluorescents that will use a cord and plug whether AC or DC, with a three prong plug and a three wire cord. The ground wire is grounded to the metal fixture. This is required by code precaution, without it you are using a high voltage appliance (on the bulb side of the ballast) that can use you as a grounding path to the sink, tub, or basement floor. Charred alternative energy users give PV a bad name.

The largest percentage of the load in a small PV system will most likely go to lighting. The efficiency of lighting is based on how many lumens of light are produced by how many watts of electricity. The second consideration is placement of lights. Light decreases in intensity by a square factor as the distance from the source increases. Thus a small light closely placed has as much usable light as a larger distant light. Hand in hand with light placement is fixture design. Some fixtures use a reflector to concentrate light in one area, some diffuse light, and some just plain waste it.

The best news for the PV owner is that DC lighting is more efficient. A 50 watt 12V DC incandescent bulb supplies the same light as a 100 watt 120V bulb. This is not magic; it is a simple case of amperage. The 50 watt 12V bulb uses 4 amps; the 100 watt 120V bulb uses .1 amp. The higher amperage bulb hits the better efficiency range of the tungsten filament. The next step up in efficiency from DC incandescents is fluorescent lighting. A 13 watt fluorescent (AC or DC) will give you the same light as the 50 watt 12V DC bulb or the 100 watt AC bulb. Fluorescent lighting is further broken down into types of light. The most common are cool white and soft white. The newest and most natural light comes from the twin tube fluorescent lights. Another type of light bulb is the quartz halogen. It is slightly more efficient than the 12V incandescent but has a bright natural light.

Take as much care and get as much help in designing your lighting system as you do in designing your PV system. Our recommendation is to wire your house up to code for AC distribution. This will give you AC receptacles and light fixtures everywhere you will ever need them, now or in the future. If you do not have an inverter, then all lighting will be on additional DC circuits. Inverters have become very efficient and reliable in the last five years. We recommend using mostly AC lighting if you own an inverter. DC wiring is an expensive duplication. We do, however, like to see one DC fixture or receptacle in each commons room. This will mean a couple of DC circuits in your home. These lights will work if an inverter breaks, or if you have the inverter turned off during a bad lightning storm. These lights can be the lights you use most if you like, not just emergency lights on the wall.

DC fluorescent lights and AC compact fluorescent lights, such as the Osram DULUX® EL* bulbs, are powered by a different ballasts than conventional AC ballasts. These very high frequency ballasts eliminate flutter and irritating 60 cycle hum. After that it is a choice of type of light. Standard fluorescent bulbs come in cool white or soft white. Many people try to avoid fluorescent lighting, and feel most comfortable with the spectrum of light from an incandescent bulb. In our business and house, fluorescents light the offices, bathroom, kitchen, cellar, and the general lighting of the other living areas. Hall lights and attic lights that are never on for any significant time period are plain old inefficient 120V incandescent lights.

The best and most efficient light yet comes from the new Osram DULUX®EL twin tube bulbs. These are 120V AC with a standard screw base. They are more compact and lighter than the Phillips, Mitsubishi, or Panasonic. They start instantly; no annoying flicker, flicker, flicker before they come on. They come in four sizes and wattages, 7W, 11W, 15W, and 20W. The trim sizes will fit in most fixtures and lamps designed to use a standard incandescent light bulb. They also have an 11W and a 15W with a reflector built on the bulb. These replace the standard 75W incandescent

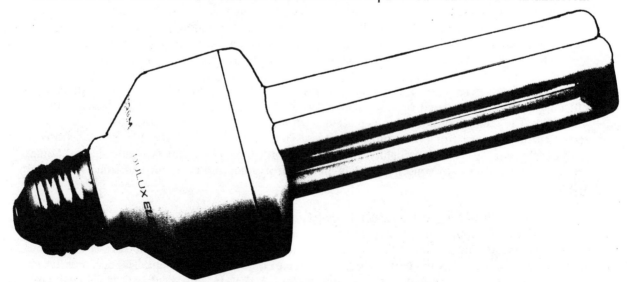

Product Info. 10-1 Osram Dulux®EL 15W Bulb Actual Size

*DULUX®EL is a registered trademark of Osram Corporation, Montgomery, NY 12549

spotlight. Their spectrum of light is very close to that of an incandescent light. The 15W bulb is rated with the same lumens as a 60W 120V incandescent bulb. However this rating is for the lumens when the bulb is first turned on. In five minutes the bulb becomes 25% brighter.

Watts	Diameter (D) (inches)	Diameter (D) (mm)	MOL (inches)	MOL (mm)	WT (oz)	WT (g)
7	2 1/4	58	5 11/16	145	4.1	116
11	2 1/4	58	5 11/16	145	4.1	116
15	2 1/4	58	6 7/8	175	4.6	131
20	2 1/4	58	8 3/16	207	5.0	142

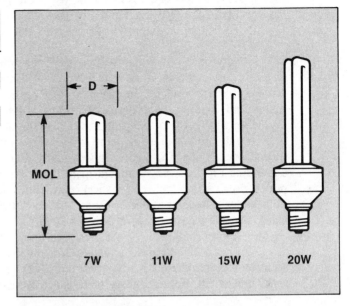

Color temperature: 2700K
Color rendering index: 82
Rated average life: 10,000 hrs.
Burning position: Universal

Watts	Base	Volts	OSRAM Lamp Reference	OSRAM Order Code	Package Quantity	Initial Lumens	Comparable Standard Incandescent
7	Medium	120	DULUX EL 7W	DB010	6	400	25W A-19
11	Medium	120	DULUX EL 11W	DB020	6	600	40W A-19
15	Medium	120	DULUX EL 15W	DB030	6	900	60W A-19
20	Medium	120	DULUX EL 20W	DB040	6	1200	75W A-19

Product Info. 10-2 Osram Dulux®EL Series Specifications

One Osram DULUX®EL 15W bulb being powered by the Trace inverter uses only 25 watts from the battery bank. This light in Chinese paper shade or a translucent fixture will light a whole room. If you use three 15W Osram lights at once, the Trace inverter will use only about 55W from the batteries. With this type of efficiency, there is little argument for DC lighting.

We now feel that the majority of the lighting powered by a large PV system with an inverter should use 120 volts AC efficient fluorescent bulbs. The remaining small percentage should be low voltage DC lighting. These fixtures should be in the commons rooms making up only two or three DC circuits. You will always have light even if the inverter is off or under repair. The decrease in lighting and wiring costs compared to an all DC lighting system can allow you to buy more modules.

A DC system in a PV home will be 12V or 24V. Twelve volt systems have been popular because there are lights and appliances available from the RV and auto technology. We much prefer a 24V system in all but the smallest PV system because the wiring is so much easier and less expensive. Any circuit on 24V can be longer with more lights. This means a simpler DC distribution. Most DC lights are available in 24V now. The extra cost is easily offset by the ease of wiring. A 12V TV, stereo, or phone machine will always run on 12V rather than 24V. Vanner®* Inc. makes a battery equalizer that allows you to get 12V from a 24V system. We discourage people from using a 24V PV system and a battery equalizer to power all 12V DC lighting. The best reason for 24V PV is that medium to large inverters run on 24V.

Every once in a while we find someone who is still using automotive brake light bulbs and cigarette lighter plugs. Bury your dinosaurs. Incandescent bulbs with conventional size, shape, and screw bases are available in 12V and 24V at 15, 25, 50, and 100 watts. Almost all conventional lamps and fixtures work perfectly at low voltage DC with a low voltage DC bulb. The only exception will be the one out of a thousand that does not have a clicking switch. Our favorite light is a twenty dollar drafting lamp with white inside the reflector, fitted with a 25 watt 24V incandescent bulb. The light is good quality and the lamp can be moved closer or farther away depending on the work. Other favorite lights are an Osram DULUX®EL 15W bulb in a Chinese paper globe with a pull chain socket suspended in the middle and an Osram DULUX®EL 15W bulb under a hanging reflector fixture (like a kitchen table fixture or an old street light reflector). These provide pleasing general lighting.

AC and DC fluorescent lights are exactly the same except for the ballast inside. An easy solution is to purchase a new or used fluorescent fixture, remove the AC ballast, and replace it with a DC ballast. DC fluorescents or ballasts are available from most PV dealers.

Photo 10-2 Mansfield Home (Courtesy of Fowler Solar Electric Inc.)

*Vanner® Inc. is a registered trademark of Vanner® Inc. Columbus, OHIO 43204

Chapter 11

System Efficiency and Appliance Efficiency

Energy is never lost in a reaction. All the energy going into a reaction results in the same amount coming out the other side. However, the energy on the output side may not all be usable. An example is an incandescent light bulb. A small amount of the electrical energy consumed to power the light bulb actually results in light, while the majority of the energy is converted to heat. PV system or PV appliance efficiency is the ratio of usable output energy divided by the total input electrical energy. This will always be less than 100%. Our goal in a PV system is to waste as little electrical energy as possible in the form of unusable energy.

PV is a costly way to produce electricity for a home. The main reason for its present use is the inaccessibility of a site to utility power. PV is not cost effective per kilowatt. If the PV remote home were powering the same loads as its counterpart in the suburbs, the investment would be great, certainly out of the range of the average independent power producer. The real way PV systems power a home is by extreme conservation of electricity. This does not necessarily result in substandard living. If done properly, a home operates quite normally. This is the challenge.

PV power is unfairly compared to utility power. We state in the introduction that PV power costs about two and a half times as much per kilowatt as grid power. This, however, is not true from a larger national perspective. There are costs other than the ten cents per kilowatt the consumer pays. At the national level we are all paying taxes to support the Nuclear Regulatory Commission, exploration for oil, and the siting of waste facilities. At the state level we pay for power line extensions and the Department of Public Utilities. We all pay for the water and air pollution. What we have is an energy that appears to be much cheaper than it is because we pay for it out of many different pockets.

In Europe, electric power is much more expensive than in the United States. The Europeans, therefore, approach electrical use more conservatively. Consequently, the appliance sellers sell more efficient appliances. We PV users must take a questioning approach to all uses and appliances. Once the home is set up, it will run quite normally.

The first year of owning a PV system is the most difficult. If people have never lived off the grid before, they have to completely rethink their orientation to energy use. They often think many components are not working properly. Ironically these same people will brag to us the next year about how well their PV systems are performing. We hope the remainder of this chapter will help people manage their PV systems before they are forced to learn by the seat of their pants.

The first place to seek efficiency in the PV home is in the PV system and how it is used. We will try to get the most available sunlight into electricity and the greatest percentage of that electricity to the appliances.

MODULES

Modules are temperature sensitive devices. As their operating temperature increases, their output decreases. Correspondingly, as their temperature decreases their output increases. Module mounting must allow for good air circulation at the rear to allow heat dissipation. Very cold ambient temperatures in winter will actually make modules outperform their ratings.

Modules absorb the most light energy when the sun's rays are perpendicular to their surface. Any PV system that has a fixed array angle should get the most from the sun when the sunshine is scarcest and the loads are greatest. This is normally winter. If the greatest loads are in the summer, then a tracker to follow the path of the sun is advisable.

Modules need strong direct sunlight. Cloudy days produce very little. Therefore when you get a sunny day you need all the sun. An exposure for the six middle hours of the day is best. If you have a tree that shades all or any of the array, realize this will have a great effect on module production. Exposure through deciduous trees with no leaves is less than 65%. A dark shadow from a close object such as a tree trunk or a pole will sap one half of the output of each module shaded!

Module output can exceed the rated output in conditions where sunlight is concentrated on the array. This happens most often in snow country. The winter adjustment is very upright. The module receives sunlight from the sun and additional reflected sunlight from the snow in front of the modules. If the array is on the roof, the sun can be reflected from the snow in front of the house. The expanse of snow in front of the array needs to be twice the height of the array from the ground. This can increase output by up to 25%.

Line losses, due to undersized wires from the modules to the charge controller, result in decreased module voltage. If severe enough, the module will simulate a lower voltage self-regulating module that tapers its output as the battery voltage increases. This will decrease array output.

CHARGE CONTROLLER

The greatest inefficiency in a PV system is that the modules are disconnected from charging the batteries a large percentage of the time. This is due to a feast or famine incompatibility of module technology in relation to battery technology. Once the batteries are mostly charged, the array output is reduced to a trickle charge. Theoretically, a diversion of the array to another load would be gravy. So far there are no great choices. The alternate load must be non-essential. Heat or hot water look good, but the amount of energy is so small in many systems that it does little. An alternate battery bank is possible if an additional charge controller is used to keep the second bank from over charging; but why not just have one large bank in the first place? Cistern water pumping is also possible.

INVERTERS

System efficiency can be increased by load management. Heavy loads should be used on sunny days when possible. The battery bank voltage is higher on a bright sunny day under charging conditions. A load will directly use the power produced by the modules without the modules first charging the batteries and losing power to the inefficiencies of battery charging. If the batteries are full on any given day, the array will be disconnected. The output for the rest of the day is wasted. If a heavy load is used during this time it is being powered by energy that could not go into the full batteries. If the load is instead powered later at night the batteries are drawn down; perhaps the next few days will be cloudy.

Photo 11-1 A PV Home (Courtesy of Fowler Solar Electric Inc.)

BATTERY BANK

We use deep cycle batteries in our PV battery banks. However, it is best to only shallow cycle the batteries. We have discussed how this contributes to longer battery life, but it also usually results in greater efficiency of the batteries. For years we have seen this happening in PV systems. The PV owner who is running his battery bank from 80% down and back up to 60% down for long periods in the winter has more problems powering the same loads as the PV owner who cycles his batteries from full to 20% down and back to full.

Theoretically, a lead acid deep cycle battery has the same efficiency of conversion of electrical energy to chemical energy and back from 100% charge down to 20% charge. As a battery is discharged, lead sulfate coats the plates. This is a normal healthy part of battery operation. However, when the sulfate stays on the plates for an extended period of time, it goes into a crystalline form that is difficult to remove in the charging process. The percentage of the plate covered with the crystalline sulfate is a non-functioning part of the plate until it is driven off.

When a battery bank continually returns to full charge, it works at optimum efficiency. In battery banks where the batteries are cycling from 80% down and up to 60% down for an extended period of time, more crystalline sulfate forms on the plates. The plates in effect become smaller. Extra electrical energy is needed to drive the sulfate off. This is wasted energy. In extreme cases the batteries become resistant to taking a charge. The owner thinks the batteries are no good. In fact the PV system cannot supply a heavy enough charge rate to drive the sulfate off. The batteries can be removed and charged up to full on a commercial charger to restore them.

Normally sulfation can be driven off the plates with an equalizing charge from the modules or a standby charger. Remember that it is also wasteful burning more gasoline than necessary to charge the batteries. The solution is to not allow the batteries to stay down. You can eliminate loads earlier before the batteries are down or run your PV/GEN hybrid more before the batteries are down. Unfortunately we see people using the backup generator charger to only bring their batteries up a small amount. The problem is that once a large battery bank is deeply discharged, it requires a very long generator run to bring them up to full.

A PV module produces more watts at a higher battery voltage. Some of this higher wattage results in more electrical energy being transferred to the battery bank than if the battery voltage were lower. This means an array puts more electrical energy into a nearly full battery bank than it does into a deeply discharged battery bank.

Batteries that are down in charge are also supplying a lower voltage. An incandescent bulb at a lower voltage puts out less light. A DC motor such as a DC water pump turns more slowly at a lower voltage and pumps less water. Fully charged batteries will increase the efficiency of these appliances.

Inverters do not lose efficiency as the battery voltage drops. However, when a battery bank is very low, a large load will pull the voltage down to a level around 11 volts. At this point the inverter efficiency falls off rapidly. This is another argument for maintaining a higher battery state of charge.

Try to manage you loads so that your batteries return to full charge at least once a month in the worst times of the year and at least once a week in other times of the year.

WIRING

A wire that is too long or too small for the amperage passing through it will result in a voltage drop. If you are powering a 12 volt DC pump at only 90% of 12 volts because of a 10% voltage drop in the wiring, then the pump will turn only 90% as fast and pump only 90% as much water. To put this into personal economic terms, if your whole home had a similar 10% voltage drop then you would need to buy a 10% larger PV system.

Most PV owners carefully design their low voltage wiring to guard against the inefficiency of a voltage drop. Your electrician, you and your home wiring book, or your well pump man designs 120 volt wiring that allows up to a 10% voltage drop. You will want the smallest voltage drop possible to your 1000 watt washing machine or your 1000 watt deep well pump. Use our voltage drop tables in the appendix to size specific high load circuits. Also check the voltage drop of any extension cords you use for large loads.

LIGHTING

We spend a whole chapter on efficient lighting. Here we would like to remind you that good lighting placement and good lamp or fixture design will get more light where you want it without using larger bulbs or turning on more lights.

PV/GEN HYBRID

A PV/GEN hybrid is not merely a PV system with a backup charger. If this were the case, then all loads would run from the batteries and all generator output would charge the batteries. This would increase the generator running time, use the batteries harder, and suffer the inefficiency of battery charging.

In an efficient PV/GEN hybrid, we want to power a heavy load whenever the generator is on. This load is supplied directly by the generator, so there is no drain from the battery bank. The battery charging component should use as much of the remaining maximum generator output as possible to charge the batteries. Thus the battery bank does not have to power a heavy load; and while the generator is powering the heavy load, the batteries are being charged.

The best home use of the PV/GEN hybrid is for washing clothes when the battery bank is down. The washer and the water pump are powered by the generator and do not drain the battery bank. This is the greatest saving. The battery charger at the same time increases the amount of electrical energy in the battery bank.

Battery chargers usually taper their amperage as the voltage of the battery bank is driven up. We would like to charge the batteries at the highest safe charging rate to minimize generator on-time. Unfortunately, a high charge rate pushes the battery voltage up more quickly.

Trace inverters have a programmable battery charger. If you have a large battery bank and it is winter, we suggest setting the taper voltage of your inverter to 15 volts, and the amperage at the maximum. Now if your batteries are down, you will get a lot of charging in one half hour to one hour of generator time. For small battery banks (less than eight 200 amp-hr six volt batteries) the batteries may be charged too hard. If your batteries normally are nearly fully charged and the generator runs for long periods of time to power other loads, the settings must be lower. Consult your dealer concerning the charge rates for your battery bank size.

PV/GEN hybrids are sometimes used in conjunction with a shop. The greatest efficiency of generator time is accomplished if large tools that must be run for a significant time are powered directly by the generator. Small tools are powered by the inverter, and the generator need not be started.

The Trace inverter uses a pulsing type of battery charger that is very effective in removing crystalline sulfate from battery plates. We have used this to save batteries that appeared to be unsalvageable.

NEW EQUIPMENT NEEDED

Bobier Electronics Inc. manufactures a device which they call a linear current booster. This device rearranges the amperage and the voltage output from a PV array such that, at times of low battery state of charge, more electrical energy goes into the batteries. In winter, when batteries are low, this can increase array charging by 20%. Unfortunately, neither Bobier Electronics Inc. nor anyone else makes a unit that matches with a thirty amp charge controller. We need one that has a maximum input of 30 amps and a maximum output of 40 amps. It should go between the charge controller and the batteries. When someone makes this device and puts it in the right package, every PV system should be retrofitted with one.

Now that we have considered the PV system in terms of efficiency, we must look at the efficient use of appliances and the choosing of efficient appliances. If you have not yet read the Chapter 16, "Understanding the Efficiency of Your PV System" please do so. You need to understand the following formula:

Watts = Volts X Amps

APPLIANCES TO AVOID

Appliances that use heating elements and that are on for more than a few minutes should be avoided. These appliances include electric water heaters, electric clothes dryers, electric heaters, dishwashers, hair dryers, and coffee makers. Other appliances that hurt the energy budget are those that use a small amount of power but are actually on for a long period of time. These include timers, AC telephone machines, circulating fans, circulating pumps, and refrigerating units.

REFRIGERATION

Refrigeration should not be provided by the standard refrigerator. Refrigerators are generally the most inefficient appliances in the conventional home. They only use 300-400 watts, but they run about twelve hours per day. Sun Frost™* refrigerators are highly efficient low voltage DC refrigerators. These can be used in the PV home. They are best in sunny parts of the country or for non-winter use. We find in New England winters that they severely tax a PV system because they are a dedicated load that must be used every day even when the PV system is low. Most PV homes will use gas refrigerators.

Freezers are in demand at any homestead. Sun Frost™ makes a freezer. It uses considerably more electricity than the Sun Frost™ refrigerator since the food must be kept at a lower temperature. Gas freezers are available for $1500 or more for a seven cubic foot model.

HEATING

Home heating should be supplied by wood stoves, passive solar, or propane space heaters. Most any furnace system will use circulating fans or circulating pumps that run a large percentage of the day, taxing the PV system in the time of year when production is least. The use of these circulating fans or pumps would necessitate a larger PV system or a generator that is run more often.

HOT WATER

Hot water can be heated by an LP tankless hot water heater, a conventional LP hot water heater, or a solar collector. Solar hot water should not use AC circulating pumps. They will use a lot of electricity. Thermosyphoning units or PV direct powered circulating systems are preferred.

*Sun Frost™ is a trademark of Sun Frost™, Arcata, CA 95521

WATER PUMPING

The most efficient water pump is a low voltage DC pump such as the Flowlight™ booster pump. This will only be useful if the well is close to the house and the water is less than 18 feet below the pump. For deeper wells and wells that are farther away from the home, we recommend a conventional high quality 120V submersible pump or a special DC submersible pump. Jet pumps are not recommended because of their low efficiency.

WASHING MACHINES

Avoid heavy duty Maytag washers or others that copy them. They have a direct drive gear box to turn the agitator. If a washer has more than one agitating speed, it is likely to be this type. Such washers overheat an inverter because the agitator applies force on the motor to try and turn it backwards each time the agitator switches direction. Basic style washing machines use a clutch mechanism that eases the change in direction forced on the motor.

The motor should be a capacitor start motor to help the inverter start the washer. If it is not a capacitor start, a capacitor can be added to the starting windings circuit to convert the split phase motor to a capacitor start motor. You will need some further instruction on this from an appliance person.

KITCHEN STOVES

Microwaves will work on inverters. They sometimes take 20% longer than the estimated manufacturers' cooking times. They use a small amount of the energy budget because they are on for short periods of time. The automatic turn on timers will be a problem.

We recommend propane stoves. If you have not yet purchased a stove, we recommend a model with pilots. The pilots on new stoves are much smaller and use much less gas than those on older stoves. Most companies carry at least one basic model with pilots; be persistent with the salesperson if you are told otherwise.

If you already have a gas stove requiring 120V AC power, you may encounter compatibility problems with the inverter. The Trace inverter may not come up to full power but instead may stay in its "sensing mode" or "idling mode" (the Trace will be quietly ticking); or it may constantly remain in "full power mode" (the Trace will be humming audibly).

If the stove does not draw enough current to bring the inverter up to power, then, of course, the stove is not receiving the power it needs to ignite the burners or to sense and maintain proper temperature. In this case, it will be necessary to turn on an incandescent lamp or other device requiring more than 6 watts of AC current before using the stove. This will provide an adequate load to switch the inverter from sensing mode (ticking) to power mode (humming). Experiment to determine whether it is necessary to leave the lamp on once the stove is actually in use. Use of the oven is likely to require the continuous current draw of the external AC light to keep the inverter at full power (that is, keep the Trace humming).

If the inverter stays on continuously and the stove is not in use (and you have made certain there is no other AC load on the system), then the stove has a constant demand for AC power which is greater then the minimum power load required by the inverter to switch from "sensing mode" to "power mode". In this case, it would be necessary to provide a switch to turn the AC power to the stove OFF when not in use and ON when needed. Another approach, although less convenient, is simply to unplug the stove when not in use.

It is essential to realize that if the inverter is left on (humming) for 24 hours per day a lot of power will be wasted and the overall efficiency of the PV system will be drastically reduced.

ELECTRONIC APPLIANCES

Electronic appliances all use low voltage DC. The majority of them are of course marketed for a 120V AC home and have a power supply inside the appliance that changes this 120V AC to the low voltage DC required. The power supply will most likely be very inefficient.

When possible, tape players, small TVs, and radios should be the automotive or RV 12V models. Some appliances such, as answering machines, typewriters, and laptop computers, will have external power supplies and will unplug at the back where a low voltage DC input will be labeled. Pay attention to positive and negative. You may be able to use these without the power supply, depending on the voltage of your PV system and the voltage of the appliance.

Larger electronic appliances, such as stereos and 20" color TVs, will have internal power supplies and must be powered by an inverter. The labeled power requirements on the rear are deceptive. Our 20" color TV is rated at 100 watts maximum. In fact it draws 55 watts. It would probably draw 100 watts at full volume and with the brightest test picture on the screen. Stereos and computers have similar requirements. Some appliances have power supplies that constantly use a small amount of electricity even when the switch is off. These may keep an inverter in the power mode and not let it go to idle mode. These appliances should be unplugged.

There are new appliances such as portable power tools and cordless telephones that are powered by batteries that are rechargeable. The charger must recharge the batteries over a long period of time at a low power. These units will use little power, but the inverter will use eight to twenty watts of its own all the time it is kept out of idle mode. These chargers should be used in the evening when the inverter is already powering other loads.

If an appliance is inadvertently left on all night, it is a waste. You can test for inverter loads left on with any AC light anywhere in the house. Turn the last light off and the inverter will go to idle. Turn the light on again and the light will come to full power with a short delay. If another appliance has been left on somewhere else, the inverter will not have returned to idle and the light will not have a delay.

Chapter 12

Installation Guidelines

Fusing is the most essential and most often overlooked part of a PV installation. Low voltage wiring will not shock a person, and therefore we are lulled into the false belief that it is harmless. We need to be reminded of the violent electrical reaction which occurs when a metal car tool accidentally connects the positive and negative terminals of an auto battery. A shorted battery instantly releases energy greater than that of an arc welder. In the case of a short by a wire that does not quickly melt, the battery can overheat from the short and explode.

A PV system has two primary sources of electricity. The array produces 50-1200 watts of power when the sun shines. The battery bank is the other source of electricity. It has the potential to dump its whole supply instantaneously to a direct short. A battery bank which stores 10,000 watt-hours of electricity can theoretically release 600,000 watts in one minute.

Any circuit from a battery needs to be fused. The closer this fuse is to the battery, the better the chance that no short can occur between the battery and the fuse. The wiring to the DC fuse box will need a 30 amp main fuse, and each individual circuit in the fuse box will need a 10 amp to 15 amp fuse. A few inverters have a high amperage DC circuit breaker in the incoming lines from the battery. If your inverter does not, you must put your own fuse or circuit breaker here. Size it for the full rated wattage output of the inverter, and not for the surge wattage. Fuses and circuit breakers are dependent on amperage. The amperage for a 1200 watt inverter running at twelve volts is 100 amps.

WATTS = AMPS x VOLTS

AMPS = WATTS / VOLTS

Amps = 1200 Watts / 12 Volts

Amps = 100

The inverter should have its own output fuse protection, often in the form of a main fuse or breaker in an AC service panel, and each circuit in the AC circuit box will have its own 15A or 20A protection.

So far, we have discussed fusing the circuits to which the battery is meant to supply power. Any charging circuit connected to the batteries also needs fusing. In these circuits the current flows into the battery under normal working conditions. If a direct short occurs in a battery charger or a charge controller or its wiring, the battery current will flow backward through the circuit until the fuse blows. When there is no fuse, the wiring, charging appliance, PV array, or whole house blows instead!

Fuses, switches and circuit breakers are rated for amperage and voltage, and are either AC, DC, or AC-DC. As a switch is disconnected, current arcs through the air while the two contacts are still near. This arcing burns the contacts. A DC rated switch or circuit breaker utilizes heavy duty contacts and a spring loaded mechanism to accelerate the disconnect movement. (This is the clicking in old style house switches.) AC current changes direction 60 times per second. Each reverse of current flow is separated by an instantaneous off phase. An AC switch or circuit breaker is lighter duty because arcing automatically halts when the AC current goes off 60 times per second.

The DC rated switch or circuit breaker is naturally more expensive. As the amperage rating increases, it gets still more expensive. The most expensive case is a switch or circuit breaker to an inverter. The current is DC and at a high amperage. One way to avoid the problem is to only use fuses. A fuse can be removed to disconnect a circuit. The inverter or appliance should be turned off before the fuse is removed so there will be no current flowing and no arcing. Some people use a motor disconnect box that is rated for AC. Fuses are sized to need and the knife switch is connected and disconnected normally when the inverter is not loaded. Any AC load should be disconnected first with its own switch or circuit breaker. In an emergency, a quick movement of the knife switch in the motor disconnect box will probably handle the DC current. We recommend paying the extra money for an AC-DC fully rated motor disconnect box. Fully rated means it can be connected or disconnected under full load. A 100 amp model will cost about $130. Any equipment must follow current electrical code and be approved by the local wiring inspector.

Getting the right fuse for the right application is complicated. Fuses are current interrupting, slow blowing, or fast acting. For DC current, you need a fuse that is designed to prohibit arcing across the void in the circuit which is created when it blows; otherwise, in the case of a direct short, the current flows regardless of the burned fuse.

Regular distribution circuit breakers are UL® listed for AC use. The majority of these are not meant for DC use and the contacts will deteriorate and fail. We use a company that tests its breakers for DC use below 50 volts. This company does not UL® list them because of the extra expense. However it does rate them for DC use in the small print of its catalog specifications. If you have a careful electrical inspector, furnish him a photocopy of the manufacturer's specifications. He does not have to accept non-UL® listed equipment, but he probably will.

A ground is a wire connected to a conductor buried in the earth. In a PV system, a ground can bleed off a lightning surge on a module mounting structure, or an inductive surge that would damage a piece of electronic equipment. A PV system must be grounded at the mounting structure, and at each individual module frame; at the negative terminal of the DC battery bank; at the inverter casing; and at the ground bus in the AC fuse box. If these parts are connected to two or more different grounds (a metal ground mount, or a ground rod, or an underground metal piping) then all grounds must be interconnected. If not, a lightning strike near one ground can send a surge through the house wiring to the other ground, damaging electronic equipment on its way.

MACRO-INSTALLATION

Find a cool, dry, ventilated area for the battery bank, inverter, and control board. Batteries should be spaced no less than one quarter of an inch apart, on a shelf or table, designed to support their weight. On the wall separating the batteries from the rest of the area, mount a 5/8 inch piece of black painted plywood with an approximate minimum size and configuration of 4'x4'. On the side of this plywood away from the batteries, screw mount the charge controller, inverter, all meters, DC fuse box, AC fuse box, and all other in line switching and fusing. This will keep all the equipment with switches and potential sparks away from the batteries. The inverter should be close enough to the main battery terminals to require no more than four foot cables. In extreme cases, this distance can be extended and the cables enlarged.

If there is any question of safety, the batteries should be vented to the outdoors. Hydrogen is lighter than air. You can make a simple venting system with polyethylene, one inch PVC piping, some duct tape, and some stiff wire. Form a pyramid shaped, polyethylene and wire hood with at least a one inch PVC pipe exiting from the apex. The pipe should then maintain a slight uphill grade until it exits via an outside wall. Once outside, the pipe should rise four to eight feet and be protected from precipitation and nests with a screen and a vent cap. Test for upward draft. Suspend the polyethylene hood six or eight inches above the batteries for air circulation. We recommend some sort of vent to the outdoors in any installation. (See Figure 12-1.)

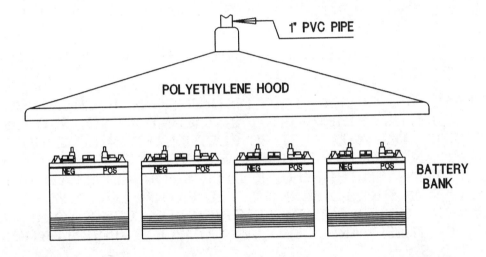

Figure 12-1 Homemade Battery Vent

Outside the building, but near the battery bank, drive an 8' or 10' ground rod. Another acceptable ground is a metal pipe that exits the cellar to run underground. If it changes to plastic pipe, forget it!

Design the module frame to hold additional future modules, or plan for additional future frames. The frame can be made of aluminum or angle iron, welded or bolted together to create a rectangle to which the modules are bolted. This frame is then hinged and attached to the mounting on the top or bottom, depending on the mounting. The other end is supported by adjustable support legs for changing the seasonal angle. In some cases you can skip the rectangle and use two angle rails; one to fasten one set of module ends, and the other to fasten the other set of ends. A third and easy option is to purchase a premade aluminum frame complete with bolts, predrilled holes for modules and mounting apparatus. This will be a fixed angle mounting system that you may then modify to make it adjustable.

Wall and roof mounts must be bolted or lag screwed to rafters, wall studs, or reinforced areas. The wind will try to blow them away. Roof mounts must be sealed to prevent leaking.

Ground mounts can be made of 4"x4" pressure treated posts, or heavy pipes, reaching four feet underground, or a lesser distance when set in concrete.

MICRO-INSTALLATION

Standard AC house wiring uses black wires for the ungrounded hot wires and white wires for the grounded neutral wires. We recommend this same system for all DC wiring. The positive ungrounded wire should be black and the negative grounded wire should be white. The other accepted method for wiring DC uses red for the positive and black for the negative. This method is used in electronics, the RV industry, and the auto industry. Inverter leads and RV equipment will normally use this system. Whether for a DC appliance or a DC system, if the two wires are red and black, then the red is positive and the black is negative. If the two wires are black and white, then the black is positive and the white is negative. Be careful and be consistent. Polarity must not be reversed to any DC appliance other than an incandescent light. When you are in doubt; test the circuit with a voltmeter first.

All wire sizes must be adequate for the amperage of the circuit, the length of run, and the overcurrent protection.

THESE ARE GENERALIZED INSTRUCTIONS. EACH INDIVIDUAL SYSTEM WILL HAVE DIFFERENT PARTS, COMPONENTS, AND MOUNTING CONDITIONS, AND MUST BE CUSTOMIZED ACCORDINGLY. AN ELECTRICIAN OR OTHER SUITABLE PROFESSIONAL SHOULD DO THE INSTALLATION AND THE SPECIFIC DESIGN FOR INSTALLATION. EACH INSTALLATION SHOULD BE INSPECTED BY THE LOCAL ELECTRICAL INSPECTOR. FOWLER SOLAR ELECTRIC INC. ACCEPTS NO RESPONSIBILITY IF THESE ARE USED AS COMPLETE INSTALLATION INSTRUCTIONS.

1. Install modules on module mounting structure. Use stainless steel bolts which are compatible with either installation of aluminum modules on an aluminum structure, or aluminum modules on a steel structure. Use stainless steel washers to separate and prevent direct contact between aluminum modules and steel structure.

2. Connect mounting structure to ground rod with 4 gauge or 6 gauge stranded wire. Where the wire travels down the building, use insulated standoffs. This ground must be bonded to all other grounding members in the PV system, the DC system, and the AC system.

3. Securely tape an opaque sheet of material over the whole array to prevent power production during installation.

4. Wire module terminals together with 10 gauge stranded wire covered with sunlight resistant insulation. For some terminal designs where the stranded wire is not clamped with a retainer plate under a screw, it will be necessary to solder or crimp on a ring terminal connector.

All positive terminal interconnects use black wire or color code.

All negative terminal interconnects use white wire or color code.

All positive to negative series interconnects use a third color coding.

12 VOLT SYSTEM (See Figure 12-2.)

Connect all positive terminals together in parallel.

Connect all negative terminals together in parallel.

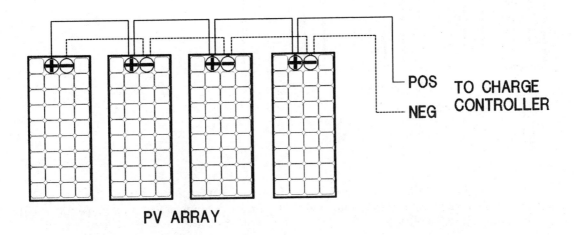

Figure 12-2 Wiring a 12V Array

24 VOLT SYSTEM (See Figure 12-3.)

Divide total number of modules into pairs. Wire each pair in series from one positive terminal to one negative terminal, effectively creating a double sized 24 volt module. Wire all the 24V double module positive terminals together in parallel. Wire all the 24V negative terminals together in parallel.

5. Estimate wire length from module array to battery bank. Calculate array peak amperage output. Look up wire length and amperage on 12V or 24V tables to determine wire size needed. A 5% voltage drop is acceptable but a 2% is better.

[Larger arrays may be broken down into four-module or eight-module subarrays. Each subarray would have its own pair of wires going to the array disconnect box. These subarray circuits may require their own circuit breakers to satisfy the *NEC*®. A subarray wire could overheat or burn if the subarray wires have a smaller rated maximum amperage than the smallest fuse or breaker between it an the battery bank. In most cases you will have a subarray wire size that is 10 gauge or larger. If so, the 30 amp main fuse between the charge controller and the battery bank will prevent the battery bank from supplying more than 30 amps through this subarray circuit if there is a short out in the subarray circuit.]

Attach white wire to closest negative terminal and black wire to positive terminal. Leave slack in the wire to accommodate frame movement during adjustment.

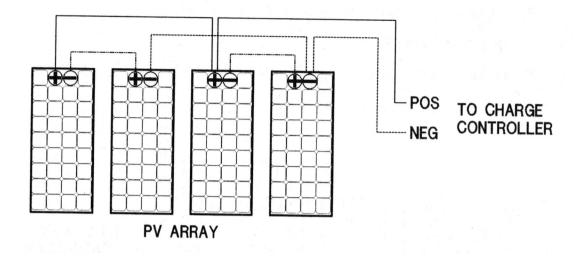

Figure 12-3 Wiring a 24V Array

Use conduit or thin walled PVC pipe to protect conventional wiring from modules to controller from sunlight or underground conditions. Or, use sunlight resistant wire or direct burial wire depending on the application.

6. Connect the wires from the array to a DC 2-pole breaker and circuit breaker box, or other suitable device. Connect the white wire(s) to the right side and the black wire(s) to the left side of the 2-pole breaker. The amperage of the breakers must be at least 25% greater than the peak output amperage of array. The peak output amperage will be the total peak amperage output of all the modules plus any seasonal increase in amperage due to reflected light from white sand or snow. In snow climates, snow reflection can increase the amperage output of the array up to 25%. Switch to off. DC fuses and a disconnect box may also be utilized. (See Figure 12-4.)

NOTE: The 1990 *National Electrical Code*® requires the installation of a DC ground fault detection device for arrays that are installed on dwellings. So far we do not know where to purchase this equipment or of its use. This will probably happen in the future.

7. 12 VOLT BATTERY BANK (See Figure 12-5.)

Divide 6V batteries into pairs. Wire each pair in series with an 8-12" 4 gauge battery cable with lug ends. One lug attaches to a negative terminal of one battery, the other end attaches to the positive terminal of the other battery. These pairs effectively become 12V batteries. Parallel remaining positive terminals with appropriate length 2 gauge battery cables. Repeat for negative terminals. Label each end of a negative terminal with white tape or paint. Test main terminals for 12 volts.

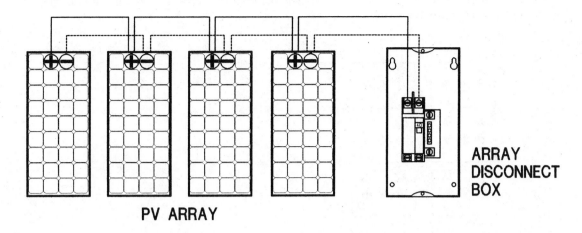

Figure 12-4 Wiring Array to Array Disconnect

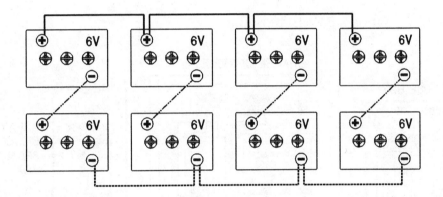

Figure 12-5 Wiring a 12V Battery Bank

<u>24 VOLT BATTERY BANK</u> (See Figure 12-6.)

Divide total number of batteries into sets of four. Wire four batteries in series with three 8-12" 4 gauge lug end battery cables. Wire quadruplets, which are now a 24V battery, in parallel with appropriate length 2 gauge battery cables. Label negative cable ends white with paint or tape. Test main terminals for 24V.

8. Connect negative terminal of battery bank to negative battery terminal of the charge controller. Connect positive terminal of battery bank to 30 amp fuse in a disconnect fuse box and then to positive battery terminal of the charge controller. If array output is 30 amps or under, use 10 gauge Romex wire or larger. (See Figure 12-7.)

9. Connect negative array terminal of the charge controller to negative side of array disconnect breaker or fuse. Connect positive array terminal of the charge controller to positive side

of array disconnect breaker or fuse. Use 10 gauge Romex wire or larger if array output is under 30 amps. Do not yet install fuses or switch on.

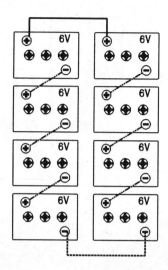

Figure 12-6 Wiring a 24V Battery Bank

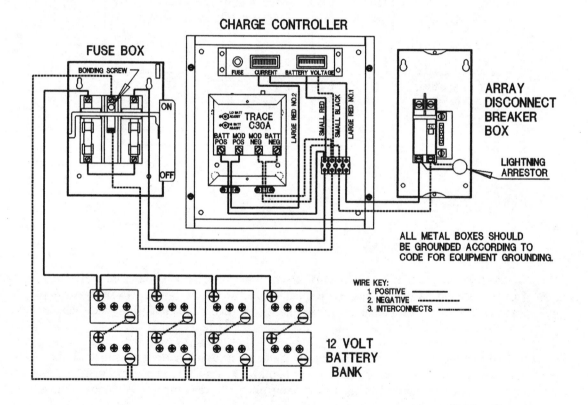

Figure 12- 7   Wiring Battery Bank to the Charge Controller

10. Use 10 gauge or larger Romex wire to connect battery bank to DC distribution main fuse box, then to DC distribution circuit box if main fusing to DC distribution is 30 amps or less. (See Figure 12-8.)

11. Connect negative inverter terminal to negative battery bank terminal with 0 gauge to 4/0 gauge 4' cable. Label each end white. Connect positive inverter terminal to positive battery bank terminal with a black 0 gauge to 4/0 gauge cable.

If the inverter has no DC circuit breaker for battery side, install a DC fuse box or circuit breaker of suitable size between battery bank and inverter. Leave circuit breaker off or fuse uninstalled. (See Figure 12-8.)

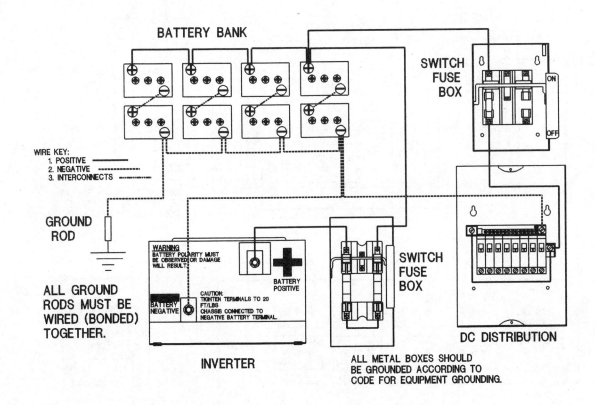

Figure 12-8 Wiring Inverter and DC Circuits to Battery Bank

NOTE: An inverter may have a red wire or terminal for positive and a black wire or terminal for negative. If you wire the inverter backwards, you will make an irreversible, expensive mistake. Check the manufacturer's installation instructions before you install your inverter.

12. Connect AC output of inverter to AC distribution panel with direct wiring or plug into inverter receptacle. (See Figure 12-9.)

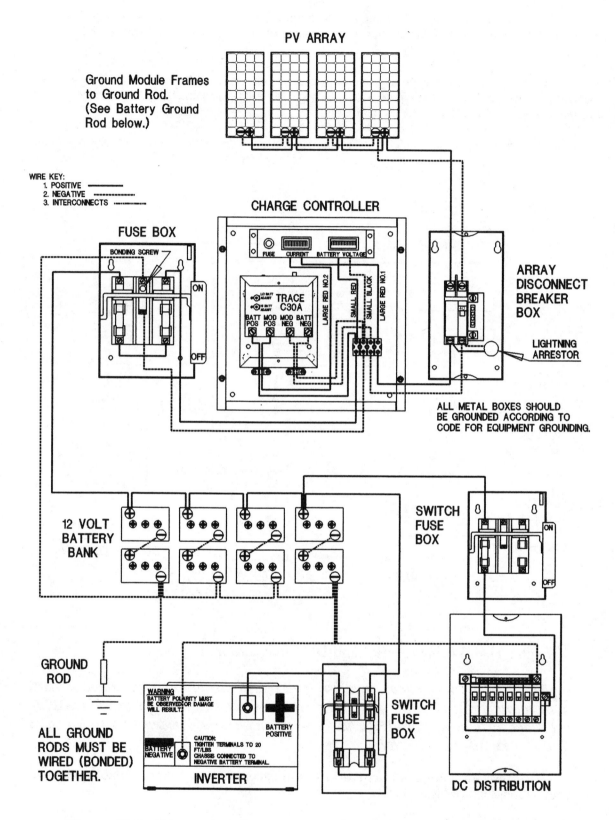

Figure 12-9   Diagram of Complete 12V PV System

13. Ground 12V negative terminal of 12V battery bank, or 24V negative terminal of 24V battery bank, to ground rod with 4 gauge stranded uninsulated copper wire.

All metal boxes such as DC switch boxes and DC breaker boxes must be grounded as they would be in an AC system. We have not attempted to cover this in these instructions. The equipment grounding method is dependent on the boxes used, and the wiring methods used. Have your DC equipment grounding design checked by the PV system designer or an electrician. See Chapter 13 for more advice on this subject.

Ground inverter casing to grounded bus bar in AC fuse box, or directly to ground if there is no grounded AC service panel. This is necessary when appliances and extension cords are fed from inverter receptacles.

Make sure AC service panel bus bar is properly grounded. If this is not grounded to the same ground as the DC system, then the two ground rods must be wired together to make one ground.

NOTE: If you are installing an inverter that has its casing bonded to the negative low voltage input terminal, then, in a DC grounded system this casing is grounded. You should not also ground the inverter casing to the AC service panel ground system. If you do you will have you a current carrying ground.

14. Remove sheet from modules. Test open voltage from array. If it is 18-20V for a 12V system or 36-40 volts for a 24 volt system, install fuses and switch on. At noon on a sunny day meters should read voltage of battery bank and peak amperage of array.

15. Install fuses or switch on circuit breaker to inverter and DC fuse distribution box. Test AC and DC with appliances.

Photo 12-1 Fresia Home (Courtesy of Fowler Solar Electric Inc.)

Chapter 13

Safe, Safer, Code

Many of us in the field of PV can remember ordering our first PV modules to power our remote cabin or home. Most likely there were no written directions, no book to be had, and no local supplier to advise us. We did the best we could making phone calls across the country and reading articles in alternative energy magazines. The result was a 12 volt system, largely based on automobile and RV technology, consisting of whatever parts were cheaply available or recommended. Some systems were safe and some were fire hazards. Today, there is information available. New PV systems can easily be installed up to the same standards as conventional grid powered electrical systems, and substandard installations can be easily upgraded.

The first priority in upgrading a PV system is safety. We want to live with a power source that will not hurt us, hurt our children, or burn our homes to the ground. The second priority is legitimacy. The installer wants systems that have a good track record and don't open him or her to unnecessary danger of liability suits. If he or she is in the business for more than just the money, then he or she wants to make the industry look good. If he or she is in it for just the money, every good system will breed new sales.

The homeowner seeks legitimacy to obtain a signed wiring permit from the local electrical inspector. With this, he or she can apply for mortgages, can purchase homeowners insurance (which can cover lightning or water or ice damage to the PV system), and can resell the property.

To complete an electrical permit for a PV system, the electrical work must be inspected and approved by the local inspector. The electrical work may or may not need to be done by a licensed electrician. The inspector must decide that an electrical system is safe. This is hard on the local inspector because he probably has never seen a PV system before. He will probably go to his copy of the *National Electrical Code*® or the state equivalent, and look up the section on Photovoltaics. He will be most secure with switching and fusing that is familiar to him. UL®* listed parts or the other registered equivalents are required, when available, over no-name products.

What is the *National Electrical Code*®? It is a guideline written by the National Fire Protection Agency offered for use in regulatory purposes. A new version is released every three years.

States or localities adopt the *National Electrical Code*® or a modified version as their guideline. The local inspector is the final interpreter and thus the person who issues the permit. The *NEC*® is meant to promote safety. Do not look to circumvent the code. It is best to take the guidelines and try to satisfy the intent of the code. In the appendix of this book you will find a

*UL® is a registered trademark of Underwriters Laboratory.

reprint of Article 690 of the 1990 edition of the *National Electrical Code®* governing the installation of PV systems.

Before we try to satisfy the *National Electrical Code®*, let us try to create a safe PV installation.

The basic small PV system that we would like to upgrade consists of an array, a charge controller, a battery, and DC wiring. Our first criterion is safety. This means overcurrent protection or fusing. When a short or draw of current above the safe level of the wiring exists, we want to disconnect the circuit. The most obvious place for this is between the battery and a load. If the wires in the wall short, we want to disconnect from the battery. We install a fuse or circuit breaker here. Most houses will have several circuits, so we will need a main fuse before the circuit fuses. We also want to disconnect the PV array from the charge controller if too much current flows from them to the batteries. This will rarely result from the module output but it can result from a surge caused by lightning. We look at the PV array and the charge controller as a duet that supplies electricity to the batteries; but current can also flow from the batteries backwards to the controller if the wires, controller, or modules short. In this case, the batteries have the potential to feed backwards in the system. So we put an overcurrent device between the batteries and the controller.

Our basic system is 12 Volts DC. We must use fusing that is safe for 12 volts DC. Don't use fuses that are rated for AC only. AC current reverses its direction of flow sixty times per second and thus shuts itself off for an instant 120 times per second. DC current flows continuously in one direction. When one breaks the circuit with a very small gap, the DC current can arc across the gap and continue the circuit path. AC current is less likely to arc because it shuts off in its normal flow pattern. A DC fuse needs to create a wider gap and to create it more quickly so the arc is stopped.

Fuses, switches, and circuit breakers are rated for how many amps of AC current they can safely interrupt or how many amps DC current they can interrupt. The same device will interrupt less DC amps than AC amps, a lot less. Practically, this means that an AC device in a DC circuit subjected to a direct short may allow the electricity to arc across the gap created and let the battery dump all of its current until it is out, or until the home is burned. Just because a fuse worked once does not mean it will interrupt in the worst case scenario.

Normal house fuses and circuit breakers interrupt 10,000 amps of AC current at 240 volts. They would be labeled 10,000 AIC AC. The value for DC would be lower even at the lower potential of 12 volts. A shorted battery dumps a whole lot of amps. You could calculate this for any battery bank. You would need the internal resistance of each battery. From this you could find the internal resistance of the whole bank. You would then use the following formula:

$$E = IR$$

$$I = E/R$$

CURRENT [of short in amps] = VOLTAGE OF BANK/INTERNAL RESISTANCE OF BANK

The basic philosophy of safe current-interrupting devices in a DC system is to use as high a rating as you can reasonably obtain. This rating should be for DC. Practically this is difficult. The normal solution is to use a top quality Class RK5 fuse. This is rated at 200,000 AIC (amps interrupting current) AC. It is in common distribution. If you speak to the engineers at the company, you will find that they are also tested for DC though they are not UL listed.

Many PV installers use regular AC rated circuit breakers of companies that have tested them as adequate for under 50 volts DC. We use them too. But these will never have the AIC necessary for use with a battery bank. If used in circuits with a good fuse between these circuit breakers and the batteries, they are much safer. The fuse is in effect a main fuse. If no fuse exists, you had better install one.

Grounding is as important to a safe PV system as fusing is. The smallest 12V systems, it can be argued, do not need grounding for their distribution circuits. In fact, a 24V system can also meet the exception of the *National Electrical Code®* for an under 50V system not needing grounding. However, once any small system is in a household, you are dealing with kitchens and bathrooms and basements that provide a greater danger of the user becoming a path to ground for current. We recommend grounding in all PV systems. It is necessary to ground for lightning protection, so you might as well ground even the small DC PV system. Once you have an inverter, there is no excuse for not having a grounded system. High voltage is deadly. Grounding is always required for 120 volt systems.

All PV systems need to have grounded frames for lightning protection. The DC system is grounded at one point on the white wire of a black and white wire system. This is normally the negative battery terminal. The rest of the distribution is grounded the same as conventional AC household wiring.

Even small PV systems often use an inverter. The AC distribution should be fused and grounded the same as any conventional household AC distribution system. An ungrounded inverter is a deadly item. This is especially true if the user is accustomed to the safety of a conventionally grounded AC system!!

The DC side of the inverter must be fused. Some inverters have a DC circuit breaker. It will not have a large enough interrupting current for a whole battery bank. Most inverters have either no fuse or a melting internal type fuse. These are adequate, if the inverter is self current limiting. Anytime there is a direct short before the circuitry, the batteries dump until the circuit is burned or the batteries are empty. The simplest solution is a catastrophic fuse bolted to the battery and then to the positive lead of the inverter. A relatively safe bubble gum and bailing wire (and not *NEC®*) solution is a 100 to 200 amp RK5 fuse with a hole drilled in each flat end. This is not recommended, but it is better than no fuse at all.

The fusing is safe, but we need one more thing. We need to have a method of disconnecting each section from every other section. This will enable installation, testing, repair, and emergency disconnecting. For now let's just say that each point of overcurrent protection needs to also provide a disconnecting means. This can be an add-on switch, a switching fuse box, or a circuit breaker. (See Figure 13-1.)

The last consideration for a safe PV system is the physical installation. The batteries need to be placed on a rack that will hold their weight. Nothing will cause shorts faster than falling batteries.

Batteries of the lead acid type give off hydrogen and oxygen gas at the high point of their charging cycle. If the battery bank is not exceptionally large the natural air changes in a conventional basement will prevent any build up of gas. Ventilation is always desirable. An enclosed small area is dangerous. Hydrogen gas is lighter than air. If there is an area above the batteries that is airtight into which the hydrogen can rise and displace the other air, you will effectively have a tank of hydrogen that is open at the bottom. Never should there be any flames or sparks right over the battery caps, the source of the gases. The simplest solution is to place all control mechanisms, that will contain the arcing DC sparks, on the outside of a partition and the battery bank on the backside.

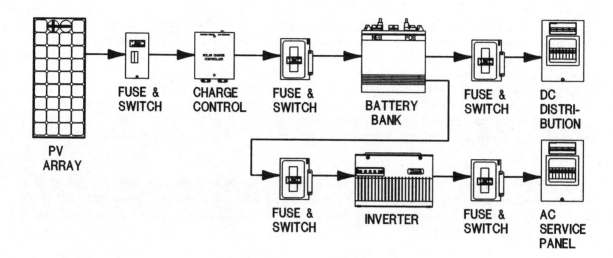

Figure 13-1   Location of Fusing and Disconnects

Gassing of batteries can be controlled at the source with catalytic hydrolyzer caps to replace the standard caps on the batteries. These add-on devices convert the hydrogen and oxygen gasses back to water before they exit the batteries.

With a little more attention to detail, this safe installation we have created can be upgraded to satisfy the *NEC*® Let's start with the battery bank.

To satisfy code, a household battery bank should be under 50 volts. It can be higher than 50 volts but then the regulations become more complex. Most stand alone systems are now 12 or 24 volts. All exposed terminals on the batteries should be insulated or protected from accidental contact during maintenance. All of us PV owners know what one tool can do when it bridges two terminals. Terminal insulation can be accomplished with special tape or heatshrink insulation. Batteries with external connections between each cell become more troublesome. Exposed battery terminals may also be removed from accidental contact by closing off the battery room with a locked vented door labeled "authorized personnel only".

The supporting structure for the batteries has to support the total battery weight. If the structure is metal, it must be coated to be a non-conductor, and it should be resistant to damage from acid. For most remote home systems, a sturdy wooden structure with acid resistant trays or a membrane on top [like polyethylene] will make an inspector happy.

To ground a PV system, first use a good grounding member. This can be a 10 foot ground rod in the earth, or a metal underground plumbing system where it exits underground. All module frames must be grounded to this point. Modules mounted on a metal frame are effectively grounded to the frame and only one point on the frame need be wired to ground. A system that grounds module to module and then to ground, needs to be wired so that, if a single module is removed, the grounding system is not interrupted. Use a 2 gauge to 6 gauge bare wire, depending on length of run.

The DC distribution system needs also to be grounded. This should be at the same ground rod or, if a different ground rod is used in another location, then the two grounds must be bonded (or wired) together with a 2 gauge to 6 gauge wire, depending on wire length. The PV system is grounded at one point and one point only. If more than one point is grounded, then the bare ground wire between these two points becomes a current carrying conductor. This can be a troublesome point, and any wiring plan should be double checked. It should also be double checked that no switch or disconnecting means can disconnect the grounding network.

Normally, the DC ground is at the negative battery terminal. Two wires with no bare ground go to the inverter and/or the DC power center. You can visualize the short run of white wire as an extension of the battery terminal that is grounded. At the inverter, the case is connected directly to the white wire (negative) terminal, thus grounding the case; or the inverter case is grounded by the bare equipment ground wire going to the AC service panel, that is grounded. The case must not be grounded to more than one of these two places. This could result in a current carrying ground.

In the DC fuse box, the white wire goes to the bus bar that is bonded to the metal box, thus grounding it. All distribution equipment beyond this point uses conventional wiring practices. The black (positive) wires from the circuits go to the circuit breakers and the white (negative) wires and bare wires go to the bus bar. Please consult your electrician about home wiring distribution.

The array wiring naturally becomes grounded when it attaches to the battery bank. However, any metal boxes of fusing, circuit breakers, or charge controllers are not grounded. They must be. They need to be grounded by a bare wire back to the single point of ground at the batteries or directly to a grounding member. (See Figure 13-2.)

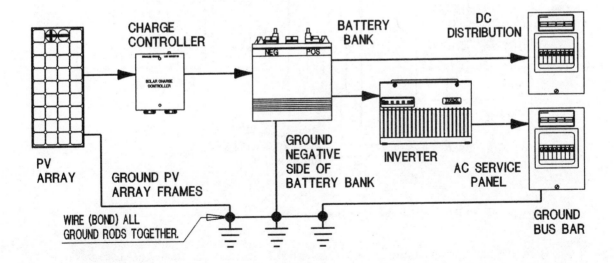

Figure 13-2   Location of Grounding Points

Now we will start at the modules and work toward the batteries. We have already satisfied the first code criterion. The frame and the module frames are connected to ground. All outside wiring has to be sunlight resistant UF wire to prevent ultraviolet deterioration. The module interconnects can be single conductor UF where only one wire is needed. Other methods of protection such as conduit accomplish the same purpose.

The 1990 *National Electrical Code®* requires the installation of a DC ground fault detection device for arrays that are installed on dwellings. So far we do not know where to purchase this equipment or of its use. This will probably happen in the future.

The positive and negative wires from the array go to an overcurrent device as soon as possible upon entering the building. This needs to be rated at no less than 125% of the maximum operating current of the array. Find this using the short circuit current ratings of the modules. We use 125% because most overcurrent devices are not meant to run at 100% for the continuous output of the array. Remember snow reflection may increase your maximum operating current by up to 25%.

The overcurrent device at this point must fuse and disconnect each of the two wires. This is also a place where current can be supplied backwards from the batteries. To satisfy code, a user cannot change a fuse and come into contact with any parts that are electrically alive. Regular fuse and switch combinations only protect in one direction. Thus, when the switch is off, only then can the box be opened, and all user accessible parts are disconnected from the electrical supply. When electricity can be supplied to a fuse and switch box from two directions, there are always electrically live parts accessible to the user. This problem can be solved by a DC circuit breaker. (See Figure 13-3.)

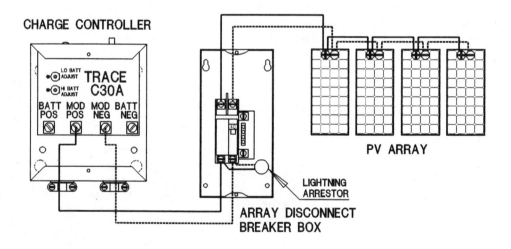

Figure 13-3 Wiring of Array Disconnect

Between the charge controller and the batteries, we need a means of disconnecting the controller from the battery bank. If we use a fuse with a higher AIC then we have protected the DC circuit breaker further out the line that only has a small AIC. Of course, we are back to the problem of power potentially from either side of the fuse and a switch on only one side. The good news is that we only need to switch the positive ungrounded wire at this point of the system. We recommend using a two pole DC rated fuse and switch box. Send the black wire in one side, put a jumper between the two output terminals, and put the battery bank wire out what would have been the other input. You end up with two fuses in series but all user accessible points are dead when the switch is open. If you have two class RK5 in series, you will have negligible voltage drop. (See Figure 13-4.)

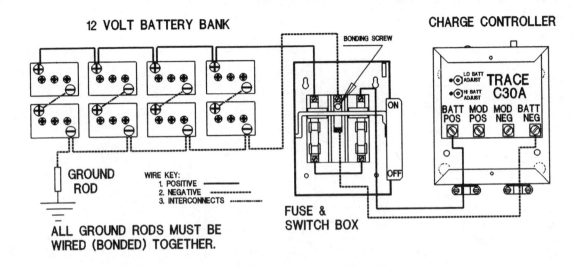

Figure 13-4 Wiring from Controller to Battery Bank

If this is too tricky for your inspector, then use a switched fuse box, fasten it shut with a padlock, and label it "serviceable by authorized personnel only".

The overcurrent protection on the DC side of the inverter can be satisfied by a switching fuse box. We need only fuse the positive leg. The inverter/charger puts us back to the same problem of current from either source and a disconnect on only one side. Use the same solutions that we used previously. (See Figure 13-5 and Figure 13-6.)

One problem we have not discussed is that many of the components we use (such as controllers, diodes and meters) are not enclosed in boxes and have exposed terminals. The terminals need to be insulated or enclosed unless they only expose the wires that are grounded and thus at ground potential.

An important goal of any PV design is to use as many UL® listed components as possible. The inspector can identify them and the design is less expensive because of mass production. To meet code and to protect all involved, these components should be exactly labeled. If the standard label applies, fine. If it is a component that is labeled AC only and you are using it for low voltage DC, label for use not to exceed 12 volts or 24 volts.

The *NEC®* requires detailed labeling of the PV system. All methods of disconnect must be labeled and grouped. The PV system must be labeled for operating current and operating voltage, and short circuit current and open circuit voltage. For example a 24V system should be labeled as 24V operating voltage and 44V open circuit voltage, and 24 amps operating current and 27 amps short circuit current. The modules themselves should be labeled satisfactorily by the company; if not, they need to be rated for open circuit voltage and short circuit current at standard conditions.

This chapter is an interpretation of the National Electrical Code® for PV homeowners to use as an educational guideline in the preliminary design of their systems. All final design decisions should be made by authorized personnel and inspected by the local inspector after installation.

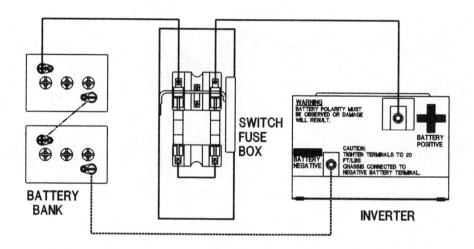

Figure 13-5   Fusing for Inverter Charger

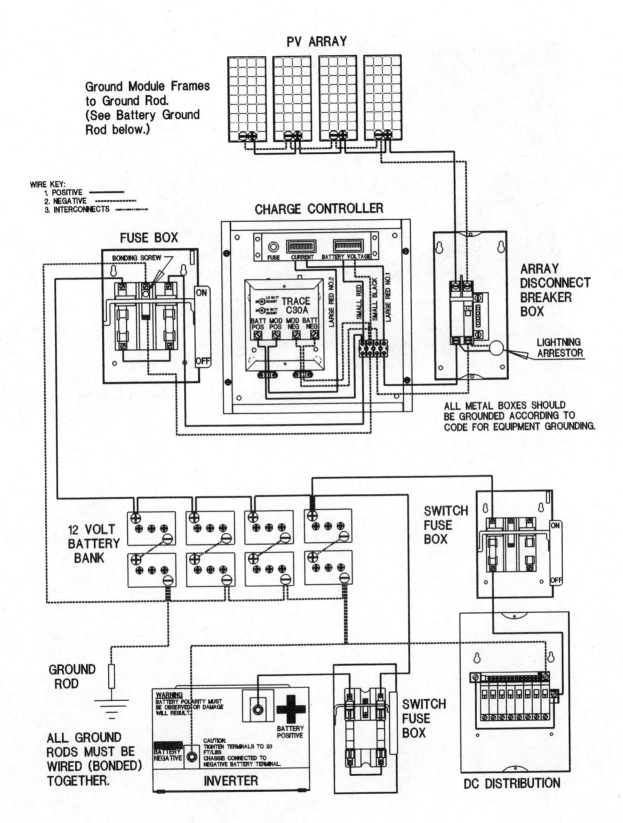

Figure 13-6   Diagram of a Whole PV Sytem

Photo 13-1 Curiosity (Courtesy of Homepower Magazine - Brian Green)

Chapter 14

Lightning Protection

Lightning is the biggest threat to remote home PV systems. These systems are low maintenance with no moving parts. However, one good lightning surge can put out a charge controller, an inverter, or the ballasts on fluorescent lights. Remote home sites are always in the open to obtain the solar exposure and often on a hill. This creates a greater exposure to lightning.

Lightning is created by a meteorological phenomenon that results in a stratification of electrical static charge. The clouds build up one charge and the earth builds up the opposite charge. Under the correct condition, a spark in the form of lightning jumps from the clouds to earth thus equalizing the potential difference. This spark is at an extremely high voltage.

A direct hit of lightning can destroy anything in its way to ground. It can blow a hole in a roof or knock down a tree. However, if it hits a conductor that can quickly dissipate the potential difference, then less harm is done. This is the principle of a lightning rod.

Direct hits are not common to a home or PV system. Probably little can be done to protect against a direct hit except to keep the hit outside of the home and thus the direct hit away from the inside components. You can protect your investment with homeowner's insurance. Insurance companies are used to paying claims on lightning damage. We have verified inverter damage on several homeowner claims. Payment was made with no hassles.

Most damage to PV systems is created by a high voltage surge from lightning strike in the near area. The high voltage surge passes through the wires in the PV system and household circuits, destroying electronic equipment in its path. Electronic equipment is controlled by components that are destroyed by any instantaneous high voltage surge.

The simplest method to protect from lightning is to isolate the damageable equipment. The computer industry sells expensive units that create a gap in the distribution of AC power to a computer so that any high voltage surge coming in from the utility company does not connect to the computer. Home PCs are most often protected from lightning by keeping the units unplugged when not in use, or when there is an electrical storm.

The most primitive way to protect a small PV system is to disconnect all equipment during an electrical storm. Both wires of the PV array are switched off; all AC loads are disconnected from the inverter to protect a high voltage surge in the AC distribution from getting to the inverter from the AC side; and if possible, the inverter is disconnected from the batteries. This is not practical for most people.

The first place to protect the PV system is outside. All modules and frames must be grounded. If all modules are bolted to a metal frame, then only the frame needs to be grounded. A 6 gauge to 2 gauge wire from the frame to a ten foot ground rod driven in the ground is the standard method. The wire at the frame must be securely bolted, or the connection will not be good enough to rapidly conduct the lightning surge to ground. Likewise, the wire should be bonded to the ground rod with a special fastener that insures good contact. Your electrical supplier can provide you with the fastener and the ground rod.

The general principles in lightning grounding are to use the largest wire with the shortest, straightest path to ground to create the least resistance and fastest dissipation of a lightning surge to the safety of ground. This will minimize the effect of a strike and minimize any high voltage surge. On a positive note, PV modules are extremely resistant to lightning damage if the frames are grounded. We've seen modules that were hit, and the only damage was to the replaceable bypass diodes ($2 each). Test studies corroborate our findings. The other protection from strikes to PV systems is the fusing between the array and the charge controller. A large surge will burn the fuses and disconnect the circuit.

As we previously stated, most often damage to a PV system results from the high voltage surge associated with a lightning strike outside. This surge can be merely induced on wires coming into the house from outside. The greatest importance here is to dissipate the surge as quickly as possible. If a surge is dissipated in zero time the surge can travel no distance down a wire. If it is not dissipated, it will travel to the appliance or inverter and do its damage. If, however, it is dissipated in a short enough time so that it just doesn't reach the appliance, then we still have protected the appliance.

To protect equipment from high voltage surges, we use devices that absorb the surge and then dissipate the surge to ground before it can reach an appliance. These devices are non-conductors of electricity at the normal operating voltage of the small PV system in which they are used. However, when the voltage rises above the standard level they become conductors. A surge suppressing device is connected between the positive wire of the incoming array wiring and ground. As an example a 24 volt PV system operates at voltages between 24 and 30 volts when charging batteries. A surge suppressing device is wired between the positive wire and ground. We choose one that is a non-conductor below 60 volts and a conductor above that. When a lightning surge hits, the voltage rises above 45 volts and all current is dumped to ground, protecting the circuit beyond this point.

These devices come from many companies in many sizes and designs. It would take another whole chapter to begin to discuss the ratings. The best advice is to consult your local PV dealer and make sure he or she has been using them in the field. In general, these devices are rated for how fast-acting they are, at what voltage they operate, and how large an amperage they can handle. The cheaper devices are only onetime protection. The better ones will act many times. A good charge controllers will already have a surge suppressor or transorb in its circuitry to protect itself.

Surge suppressors should be installed in every circuit entering or exiting the building as close to the outside as possible. In effect, the PV system is isolated from any potential path of lightning entry. The most obvious paths are the wires from the PV array. A surge suppressor is connected in the combiner box or the array disconnect box; whichever is closest to the outside. This will give added protection even if the controller has its own. The device should be connected from the positive wire to ground by the most direct means with a wire as large as or larger than the positive wire. A second suppressor should go from the negative wire to ground in an ungrounded system. If the PV system has the negative wire grounded, as recommended in this book, then the best method is to connect the suppressor between the positive and negative wires. A surge will be bled to ground by the grounded negative wire.

For good lightning protection, the DC system should be grounded at the negative terminal of the battery bank. This means that the system is by design connected to ground and therefore capable of dissipating surges. Another often overlooked method of isolation is battery bank wiring. Any battery is in effect a surge suppressor between the positive and negative wires connected to it. A high voltage surge is buffered by the voltage sinking effect of the battery. A good wiring method is to bring in the wires from the charge controller to a set of terminals at one end of the battery bank and take out the wires to the loads and inverter at the other end of the battery bank on the parallel positive and negative terminals. Thus surges are absorbed by the batteries whenever possible.

Though not common in remote homes, if a DC circuit exits the building to a garage or barn via an overhead or underground method, then this circuit must have a protection device just before it exits the building because it is a potential entrance path for lightning surges.

The inverter must be protected on the AC side even more than on the DC side. Most units will have a surge suppressor to protect their circuitry from high voltage surges created by reactive loads. The Trace inverters have been the best in our experience for resistance to lightning. We've rarely had one fail from lightning. The internal suppressors are similar to the ones we have been describing, but they are higher voltage units for a 120 volt system. They are zenir diodes.

Many of the problems we have with components in PV systems stem from the fact that they are new and not widely distributed. Often the game is to find what equipment is in general distribution. For AC 120/240V lightning protection there is common and thus less expensive protection. These devices are large and will protect the home from the large surges associated with lightning strikes to the utility poles of a regular grid system. You need to be an electrical engineer and have field experience with the specific model. Only a few of these devices act fast enough. The others will crowbar a large strike and protect your home but not before some of the surge has destroyed your electronic equipment. We use one that is large, inexpensive, and fast acting. The price is only $27.

A lightning suppressor's ability to act quickly is increased if the device is as far as possible from the device it is protecting. Simply, it takes time for the surge to travel the distance down the wire to the device. These distances are relevant values even at the speed at which electricity travels. Thus AC protection must also be installed at the closest point of entry to the building. Normally the device is wired in the AC distribution panel between the hot main feed and the grounding bus bar. Two overlooked points of entry for lightning surges are the wiring from the generator to the inverter/charger and from the deep well pump. The generator wiring should use the same device as the main AC distribution panel between the generator and the inverter/charger. (See Figure 14-1.)

In our experience, deep well pumps are the biggest culprits in lightning damage to inverters. It is the most common point of entry for lightning surges because the metal casing and underground wires are capable of having a current induced upon them. We have watched this high voltage surge arc across the relay in the pressure switch with every lightning strike. The solution is a lightning suppressor like that used at the main distribution panel except that it is a three phase model that has three hot wires in and one neutral wire to ground. Each of the three hot wires is connected to one of the three wires going out the building to the well. This device will also help protect the well pump from lightning damage. Unfortunately, many well service people think that well pump damage comes from lightning strikes to the utilities that come through the house and out to the pump. The engineers at the well pump manufacturers are better informed.

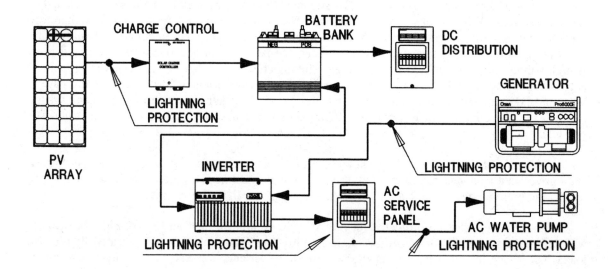

Figure 14-1   Locations for Lightning Protection

The AC system in any legal and safe home electrical system is, of course, grounded to *National Electrical Code®* specifications as discussed in another chapter. For your own safety and for good lightning protection, it is necessary to ground the case of the inverter. Another reminder, if grounding is accomplished at more than one grounding rod, pipe, or other point then all of the grounding points most be bonded together with a large enough conductor to serve the distance between them.

The information in this chapter is supplementary information to the National Electrical Code® and the local or state electrical and building codes. We expect that all wiring will be done by qualified personnel and will be inspected by your local inspector.

Chapter 15

Maintenance of a PV System

A PV system requires very little maintenance. The best care is preventive medicine. Batteries and electronic equipment should be kept clean, dry, and free of dirt and dust. Most of this is provided for with proper initial installation of the components.

Service of the remote home PV system is provided by the owner. Though we often state that service should be performed by a qualified professional, we know that, like it or not, the service will most likely be performed by the owner with telephone backup from the dealer or factory. However, if you, the owner, are over your head; please call an electrician.

In the event of a component failure, that component will have to be returned to the factory for repair. Neither you nor your electrician will repair a battery, an inverter, or a charge controller. What you will need to do is isolate the problem. We have few component failures in our systems because PV equipment is good. However, we have many calls from owners who suspect that something has failed.

When you suspect a problem, there are a few basic checks you can perform. First, decide if the new condition started soon after you made a change in the system or the system wiring. For example, have you just hard wired the generator to the inverter and now the inverter is not working properly? In this case, check the new work for errors.

If no new work has been done, check all the fuses and circuit breakers. If you do not have a continuity tester to test the fuses, you may install a new spare fuse that you are certain is healthy. Often people forget that in their PV/GEN hybrid, the generator may have circuit breakers on its distribution panel. Next look for loose or disconnected wires. If you are experienced in the use of a voltmeter, trace the supply of electricity from the source to the malfunction to see if you can find the problem. If you do not find the problem, then call your dealer or the factory. Have notes on all conditions or checks you have made with you at the telephone to help in the long distance troubleshooting.

The remote home PV systems currently in use are simple and basic with few automatic features. Cost prohibits automatic backup charging generators, fancy watt-hr meters, or high tech bells and whistles. Instead, the homeowner monitors and maintains his or her own system. We assume that every PV system has a voltmeter to measure battery voltage and an ammeter to measure the amps coming in from the array. The first place to start monitoring and maintaining your PV system is at the battery bank.

The deep cycle lead acid battery contains sulfuric acid as the electrolyte. Under normal charging rates, electricity, in the process called electrolysis, splits water molecules, releasing hydrogen gas and oxygen gas. These gases bubble from the electrolyte and escape through the holes in the vent caps. If these gasses are allowed to accumulate in a confined area they will reach a percent content in the air at which a spark or flame will ignite the hydrogen.

NO WORK SHOULD BE DONE ON OR NEAR THE BATTERY BANK WHILE THE BATTERIES ARE CHARGING AND GASSING. THE CHARGING SOURCE SHOULD BE TURNED OFF AND THE BATTERIES ALLOWED TO DISSIPATE THE GAS .

The water that changes into hydrogen and oxygen is lost to the battery. Every few months the electrolyte must be topped off to replenish this loss. Use only distilled water and a clean funnel. Distilled water is sold by the gallon in drugstores. Any spilled liquid or foreign matter should be cleaned from the battery top surface. Follow the battery manufacturer's warnings about acid dangers to eyes and skin when adding water.

With the advent of new low gassing charge controllers, many people will upgrade to a new charge controller that reduces the battery gassing, the dangers associated with gassing, and the maintenance of adding distilled water to the batteries.

When any battery terminal shows any sign of corrosion, the wires should be removed and all surfaces wire brushed. Disconnect the charge controller, inverter, and load line before disconnecting wires between the batteries. A new anticorrosion ring should be added to the terminal.

A deep cycle battery should not be discharged below its last twenty percent of charge. Though it may survive a complete discharge, this will decrease its life. This same battery needs to reach a full charge at least once every four to six weeks to maintain its full rated storage capacity. In addition, the battery bank may require a short period of overcharging once every few months to equalize the specific gravities in each of the cells in the battery bank. The imbalance normally occurs only if the batteries have not been completely charged often enough. The equalization can be performed manually by bypassing the charge controller. The Trace C-30A has a switch to equalize, thus no disconnecting of wires is necessary. However, most charge controllers manage to equalize the cells during normal operation.

To equalize the cells in a battery, all of the cells of the battery are charged until the cells that are of a lower specific gravity gradually come up in charge to the same level as the fully charged cells. The charge controller is taken out of the circuit, or defeated, so that the modules can charge the batteries at a higher voltage than is normally possible. This means the modules will not disconnect automatically during equalization. It is a planned, manual overcharge. The process of equalizing the cells of the battery bank requires careful monitoring due to the potential safety problems of greater battery gassing. The system should not be left unattended when the batteries are being over charged and are gassing.

The voltmeter on a charge controller measures the voltage of the battery bank. During daylight, the module array is supplying current to the battery bank for storage. The supply of current excites the battery voltage to a higher level. A fully charged deep cycle battery at rest (slight variations occur with each manufacturer) has a voltage of 12.6 V. The same battery reaches full charge at 14.5V, while it is actively being charged at the normal charge rate in a PV system.

A standard analog or needle type meter is not always accurate. Since we use these as the full to empty gauge for battery level of charge, we suggest that you compare the reading of your analog meter to a more accurate digital voltmeter. If you determine, for example, that your analog meter reads .2 volt low, you should record this next to the meter for reference.

To test the voltage of a battery at rest, wait several hours after sundown and read the voltmeter. The excitation voltage or surface charge can be bled off more quickly if an appliance is attached for 15-30 minutes. Disconnect the appliance and read the voltmeter. During the powering of a heavy load, the voltage will read less than normal, but will quickly return when the load is disconnected. The following chart equates percent charge in a battery bank to voltage and specific gravity. This chart is for a battery with a specific gravity of 1.260, like a Trojan battery, at full charge. A battery with a specific gravity of 1.250 at full charge, like the Exide GC-4, will show slightly lower values for voltage and specific gravity than in this table.

Percent Charge	Voltage	Hydrometer Reading
100%	12.60	1.260
75	12.35	1.220
50	12.15	1.185
25	11.95	1.150
discharged	11.85	1.120

A hydrometer measures battery charge more accurately, though less conveniently than a voltmeter. The majority of the time the specific gravity will not need to be taken. The PV system will be operating in a range familiar to the user. If the user suspects that the battery bank is greatly discharged, then the more accurate hydrometer reading should be used to verify the voltage reading.

We recommend that a hydrometer reading be taken at least every three months as a periodic maintenance procedure. If the cells show a variation of greater than .02 on the specific gravity reading when the batteries are fully charged, then the battery bank needs to be equalized.

Another battery check utilizes the voltmeter of a digital multimeter. Test the voltage of each individual battery in the bank. If all the batteries are the same age and a proper override equalization charge has been applied, then each battery will read within .05 of a volt. This test should be performed when the batteries are at rest.

Some PV owners, despite all warnings about charging the batteries to full charge periodically, either don't or can't. This allows the normal coating of lead sulfate on the battery plates to form large crystals, referred to as sulfation. This sulfation sets up a resistance to accepting a charge. The result is a cell that rises in voltage to a point where the charge controller shuts off but the specific gravity is not up to a full reading and the cell is not charged. This cell must be charged harder or at a higher voltage cutoff point to drive the sulfation from the surface. Under these overcharge conditions, the cell will gradually charge and reach a full specific gravity.

In the above situation of sulfated plates, the whole battery bank may be resistant to accepting a charge, but more likely a few cells will lag behind the rest. This is what will show up as unequal cells in a periodic test of specific gravity of every cell. An equalization of the battery bank at a higher voltage will overcharge the fully charged cells while it gradually brings the lower cells up to full charge.

A hydrometer is dependent on temperature. Always buy one with tempera-ture compensation. WHEN USING A HYDROMETER, FOLLOW ALL BAT-TERY AND HYDROMETER MANUFACTURERS' CAUTIONS CONCERN-ING SAFETY. Battery acid is dangerous, especially to your eyes. The float will indicate the specific gravity of the electrolyte. A clean hydrometer will prevent contamination of the electrolyte with any foreign material.

Every PV owner should own a digital multimeter which reads ohms, volts, and amps for both AC and DC circuits. One of the more recent digital multimeters will read up to 10 amps. This tool can do the majority of testing needed in a PV system. The voltmeter may be used to test the battery voltage or to calibrate the analog voltmeter. For the more experienced, the voltmeter may be used to trace a circuit and find the interruption in the circuit. The ohmmeter may be used to test the continuity of a fuse, and the ammeter may be used to test the output of a module or the current draw of a load.

We recommend reading the meter manual and following the manufacturer's recommenda-tions for use. Testing electricity can be as deadly as any other work on an electrical system. Furthermore, each meter is different; improper use may result in meter damage.

An adjustable array may be readjusted with the change of the path of the sun in different seasons. The array, optimally, should be perpendicular to the sun's rays at noon time. (See Figure 15-1.) The angle of the sun above the horizon in any season and at each latitude is available in the appendix. The perfect angle between the array and the horizon, on any given day, will be 90 degrees minus the angle of the sun above the horizon.

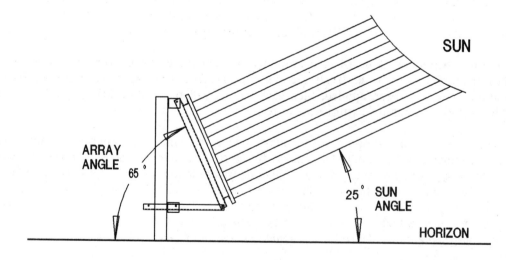

Figure 15-1 Optimum Array Angle

For a simpler method, place an approximately one foot long piece of a 2"x4" stud with a square cut end on the module surface with the square end in contact with the glass. Use this as a sundial at noon time. When the module angle is perpendicular to the sun, no shadow will be cast above or below the 2"x4" stud. (See Figure 15-2.)

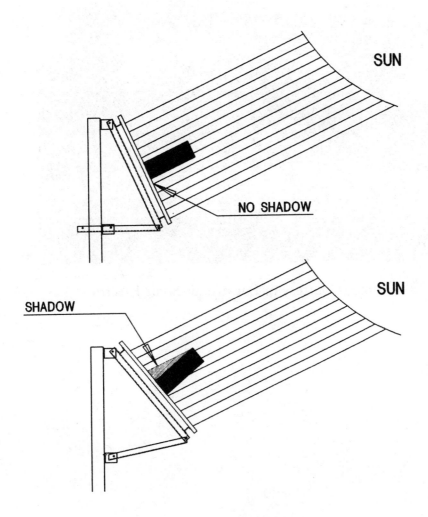

Figure 15-2   Simple Method of Array Angle Adjustment

Photo 15-1   PV Home (Courtesy of Backwoods Solar Electric)

Chapter 16

Understanding the Electricity for Your PV System

An elementary understanding of electricity is essential to any remote site home owner. It is necessary in the design, use, and maintenance of a system, and in the selection of efficient appliances. The simplest familiarity with electrical theory allows an owner to monitor, trouble-shoot, and design a photovoltaic system.

The basic blocks of electricity are volts, amperes, and ohms. Electricity can be explained most easily when it is compared to a pressurized home water system. Volts are analogous to water pressure that forces water through the piping. Amperes are analogous to the amount of water which passes through the pipes. And ohms are the resistance to the water flow that occurs when a nozzle is attached to the end of a hose.

ELECTRICAL THEORY

VOLTAGE (in volts) [V] is the potential difference in a circuit which makes current flow.

CURRENT (in amperes) [I] is the amount of electricity flowing past a point.

RESISTANCE (in ohms) [R] is the opposition to flow.

The three basic units of electricity are algebraically related as follows:

Voltage = Current X Resistance
Volts = Amps X Ohms
$V = I\,R$

This formula can also be expressed as:

$I = V/R$ or $R = V/I$

Electrical power is the rate at which electrical energy is produced or used. Power is measured in watts and is the product of the voltage and the current.

Power (in watts) [P] = Voltage X Current
Watts = Volts X Amps
P = V I

.For example, a light bulb uses 25 watts of power, or a photovoltaic module produces 40 watts of power.

Electrical energy converted to some other form of energy in a circuit is the product of power and time.

Electrical energy (in watt-hours) = Power (in watts) X Time
Electrical energy = Voltage X Current X Time
Watt-hours = Volts X Amp-hours

If the light bulb above uses 25 watts of power for one hour, then the light bulb uses 25 watt-hrs of electrical energy. A 40 watt PV module which is sunlit for one hour will produce 40 watt-hrs of electrical energy. If this is still confusing just use the following equations of units.

Power: Watts = Volts X Amps

Electrical energy: Watt-hrs = Volts X Amp-hrs

Problem 1) How much current does a 12V, 25 watt light bulb draw?

Watts = Volts X Amps
Amps = Watts / Volts
Amps = 25 watts / 12V
Amps = 2.08

Problem 2) If a motor draws 6 amps at 120V, how many watts of power does it use?

Watts = Volts X Amps
Watts = 120 volts X 6 amps
Watts = 720 watts

Problem 3) A 6 Volt battery stores 200 amp-hours. How much wattage does it store?

Watt-hrs = Volts X Amp-hrs
Watt-hrs = 6 volts X 200 amp-hrs
Watt-hrs = 1200 watt-hrs

In a simple circuit, a power source such as a battery or a PV module provides voltage, a wire conductor carries the current, and a load offers resistance. The resistance uses watts of electricity to produce mechanical energy in a motor, heat in a toaster, or light from a bulb. (See Figure 16-1.) The current flows from the battery, through the load, and back to the battery. The load and the power supply are each connected in series in the circuit. If a second battery is added to the first circuit, it may be added in series so that the current flows in one continuous loop. (See Figure 16-2.) If anything is removed in a series circuit, the circuit is broken. Or, the battery or module may be added in parallel where it adds another side circuit. If one power source is removed, only its own sub-circuit is affected. (See Figure 16-3.)

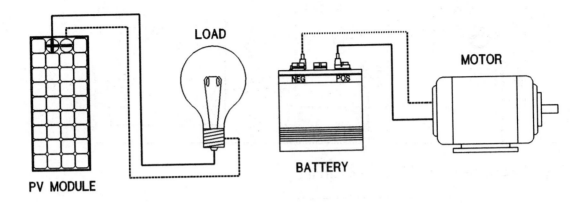

Figure 16-1 A Simple Circuit

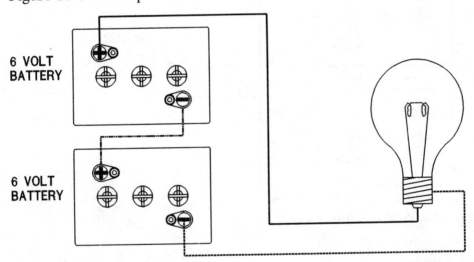

Figure 16-2   A 12V Series Circuit

SERIES: When two modules or batteries are connected in series, one's positive terminal is connected to the other's negative terminal. The current remains the same, and the voltage is additive. (See Figure 16-4.)

Example: Two 200 amp-hr 6V batteries in series store 200 amp- hrs at 12V. Each battery stores 1200 watt-hrs (6V X 200 amp-hrs). Two in series store 2400 watt-hrs (12V X 200 amp-hrs).

Example: Two modules connected in series produce 2.5 amps at 24V. One module produces 2.5 amps at 12V, or 30 watts. Two modules in series produce 2.5 amps at 24V, or 60 watts.

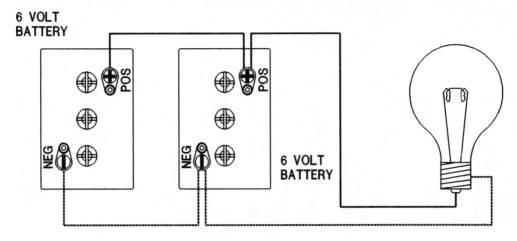

Figure 16-3   A 6V Parallel Circuit

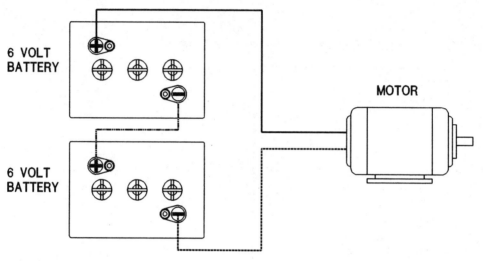

Figure 16-4 A 12V Series Circuit

PARALLEL: When two modules or batteries are connected in parallel, positive terminals are connected together and negative terminals are connected together. The voltage remains the same and the current is additive. (See Figure 16-5.)

Example: Two 200 amp-hr, 6V batteries connected in parallel store 400 amp-hrs at 6 volts. They store 2400 watt-hrs (400 amp-hrs at 6V). Note that two 200 amp-hr batteries store the same amount of electrical energy whether they are in series or parallel.

Example: Two 12V, 2.5 amp PV modules connected in parallel, produce 5 amps at 12V. One module produces 30 watts (2.5 amps X 12V). These two modules in parallel produce 60 watts (5 amps X 12V) as do two 30 watt modules in series.

Power sources may be connected in both series and parallel. This is often necessary in PV arrays or battery banks to achieve proper supply at the proper voltage. (See Figure 16-6.)

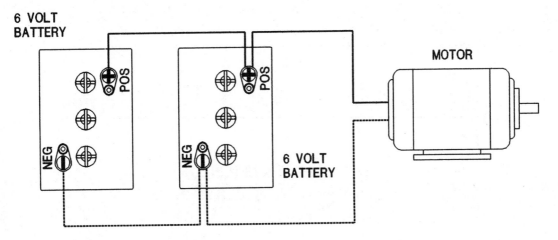

Figure 16-5   A 6V Parallel Circuit

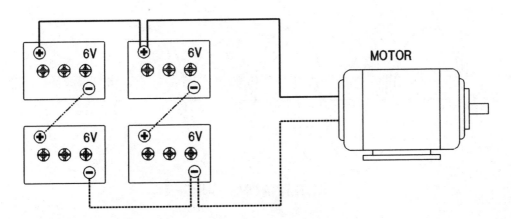

Figure 16-6   A 12V Series and Parallel Circuit

This example wires two modules in series to produce 2.5 amps at 24V and parallels this with a second pair. The two parallel pairs produce 5 amps at 24V.

A battery or a module produces direct current or DC. Current flows in one direction at a constant rate from the positive terminal to the negative terminal. A 25 watt 12V light bulb draws 2.1 amps continuously. Graphically this is represented in Figure 16-7.

Conventional household electricity is produced by the power company using a mechanical rotary generator. This generator produces alternating current, AC. During one cycle, this current is off, then on in a positive direction, then off, then on in the negative direction. The transmission from off to full on is gradual and results in a curve. If we take our same example of a 25 watt bulb at 12 volts and 2.1 amps, but make it alternating current; the graph of the current flow will look quite different. (See Figure 16-8.)

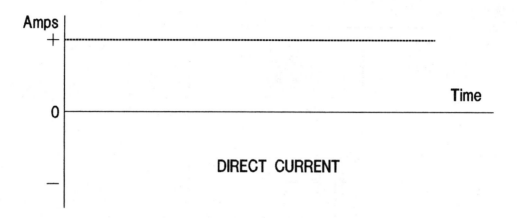

Figure 16-7   Direct Current

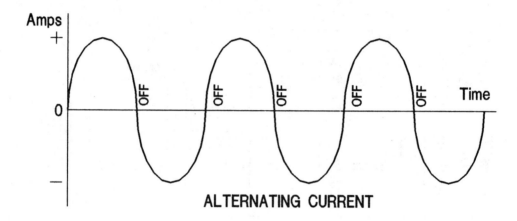

Figure 16-8   Alternating Current

One positive and one negative flow constitutes one cycle. This is followed by another and another. The timing, which is determined by the generator, must be exact. In the United States, this is 60 cycles/sec., or Hz.

Alternating current is not the natural flow pattern of current. AC current is current that is constantly being switched on and off and changed in direction from positive to negative. AC comes in this packaging because it is a product of a specific rotating generator. Unlike DC electricity, AC can be put through a transformer and the voltage can be increased or decreased. This is the reason for its use in power lines. The power company can transmit power most efficiently at high voltage, and just before reaching a house, use a step-down transformer to transform this electricity to 240V and 120V.

The previously mentioned 25 watt 12V bulb works equally well on AC or DC. It is a resistive load and does not need the current to flow any special way. An AC-DC, or universal motor, as the name implies, can use either form of current. Other motors and electrical switches are designed specifically for only AC or for only DC.

Efficiency compares the amount of energy put into an appliance with the amount of useful energy the appliance produces. A heating element in an electric space heater is nearly 100% efficient. Nearly all of the electrical energy put into the heating element is changed to heat to warm the room. An incandescent light bulb is extremely inefficient; most of the electrical energy is used to produce unwanted heat. Only a small amount is converted to light. A motor converts electrical energy to mechanical energy. The heat given off by any motor is indicative of electrical energy being lost in the form of heat.

A heating element in a space heater is a conductor of electricity, just good enough to allow electricity to pass, and poor enough to consume electricity and convert it to heat. If it were a poorer conductor it would stop the flow of electricity and be considered an insulator, like the rubber insulation which encloses a lamp cord. The wire carrying current to the space heater is a good conductor because it does not resist the flow of electrical energy and create heat before it reaches the element in the heater.

No matter how good a conductor is, it still loses some electrical energy. A lamp cord 6 ft. long loses an infinitesimal amount of energy. If the lamp cord were 600 ft., it would have unacceptable losses. This loss of energy is expressed in loss of voltage, or line loss.

An unduly long wire run in an extension cord can cause low voltage at the appliance. Voltage loss is created when a wire is too small or too long, or when the amperage required by the load is too large. To compensate for voltage loss a wire can be enlarged, the circuit can be shortened, or the load can be decreased. Again this is analogous to water. When a hose is too long and too small, there is little pressure at the nozzle.

Line loss results in a voltage drop at the far end of a wire. It depends on the size of the wire, the length of the wire, and the size of the load. It is also dependent on the voltage of the system itself. A low voltage system, like a low pressure water system, has little push to send the current down the wire. When the current is increased it has even more trouble pushing it.

It turns out that the percent of line loss is directly proportional to the length of a wire.

PERCENT LINE LOSS = (SOME CONSTANT VALUE) x LENGTH OF WIRE

For example, if you double the length of your wire while keeping the wire size (gauge), amperage, and battery voltage constant, your percent voltage loss at the load will double.

EXAMPLE

Assume you have a load of 15 amps at 24V delivered from a battery bank by a #4 wire 60 feet long (each way). From wire loss tables you will see that this will result in a 2% voltage drop at the load. Now if you increase the wire length by 2.5 times to 150 feet (each way) you will now have a 5% voltage drop at the load, a 2.5 times increase in percent loss.

If, on the other hand, you are designing a circuit with a fixed voltage loss, 5% for example, then for any given wire size the maximum wire length is going to be directly proportional to the source voltage and inversely proportional to the amperage.

MAXIMUM WIRE LENGTH = (SOME CONSTANT VALUE) x VOLTAGE / AMPERAGE

In practice this means:

1. If you keep wire size, percent loss and amperage the same, when you double the voltage, the maximum wire length will double.

2. If you keep wire size, percent loss and voltage the same, when you double the amperage going through the circuit the maximum wire length is cut in half.

When designing a circuit usually your power requirement is fixed. For example, you may wish to light a study area with an incandescent lamp of 60 watts. Whether it is supplied at 12 volts, 24 volts or 120 volts is not important to the user; only the light equivalent of 60 watts is of concern.

When you calculate how long the wires can be for this circuit, the effect of the voltage of the system becomes very clear. If you wish to use #14 wire and want a voltage drop of no more than 5%, you get the following results:

For the 12 volt system, maximum wire length to the lamp from the source is 21 feet.

For the 24 volt system, 84 feet (increasing voltage by 2 times increases the maximum wire length by 4 times).

For the 120 volt system, 2100 feet (increasing voltage by 10 times increases the maximum wire length by 100 times)!

Appendix

Table of Contents

Glossary

AC - Alternating Current. Electricity which rapidly reverses its direction of flow. One forward and one reverse flow constitute one cycle. In the USA the standard is 60 cycles/second.

AMP or AMPERE - The unit of the current flow in a circuit, specifically how many electrons flow past a given point each second.

AMPERE HOUR - A constant current flow of one ampere for one hour of time. This commonly used measure of battery storage capacity, multiplied by the battery voltage, will give the total electrical energy stored in watt-hrs.

ARRAY - A group of photovoltaic modules electrically connected together.

BATTERY BANK - A group of electrical storage batteries electrically connected together.

BATTERY CAPACITY - The total amount of electrical energy, in watt-hrs, available from a battery .

BATTERY CELL - The basic structural unit of a battery. For example, a 6 volt battery is made up of three 2 volt cells.

BATTERY CYCLE LIFE - The number of times a battery can be discharged then recharged again. The number of cycles is dependent upon the depth of discharge of each cycle.

CHARGE CONTROLLER - A device used to regulate the charging of a battery bank. Its function is to prevent overcharging of the battery bank while still allowing it to reach full charge.

CHARGE RATE - The battery charging current, in amperes, applied to a battery bank.

CURRENT - The rate of flow of electrons in a circuit, measured in amperes.

DC - Direct Current, the flow of electricity through a conductor from positive to negative. Batteries and PV modules produce DC current.

ELECTROLYTE - A chemical, such as sulfuric acid, that electrically and chemically interconnects the positive and negative plates in a storage battery.

EQUALIZATION - The process of recharging all cells of a battery bank to full charge.

EQUALIZING CHARGE - A charge at a voltage higher than that which is normally used in battery charging; the purpose of which is to bring all the battery cells up to the same specific gravity of full charge.

GASSING - Hydrogen and oxygen released when water molecules are split as lead acid batteries are brought up to full charge. The loss of water from the electrolyte solution requires the periodic addition of distilled water to lead acid batteries.

GRID - The network of utility power distribution.

HERTZ or HZ - One cycle per second. The standard for utility AC electricity in the United States is 60 HZ, or 60 cycles/second.

INSOLATION - The amount of solar energy available per area.

INVERTER - A device which converts low voltage DC electricity to 120 volt AC electricity.

KILOWATT - 1000 watts.

KILOWATT HOUR - A measure of electrical energy, equal to a kilowatt of power generated or consumed for a period of one hour.

LINE LOSS - The loss of electrical energy due to the resistance of the length of wire in a circuit.

LOAD - The appliance or appliances that consume electrical energy.

OPEN-CIRCUIT VOLTAGE - The voltage between the positive and negative terminals of a PV module when the module is not connected to any load or battery.

OVERCHARGE - The charging of a battery beyond full charge.

PANEL - A colloquial term for a PV Module.

PEAK SUN HOURS - 1000 watt-hrs per square meter for one hour. A unit of measure of the solar energy available at noon on a clear day at the earth's surface.

PHOTOVOLTAIC or PV - The conversion of solar energy directly to electrical energy.

SELF-DISCHARGE RATE - The rate at which a battery will gradually decrease its state of charge even though no loads exist.

SELF-REGULATING MODULE - A PV module which is designed to decrease its output as a battery approaches full charge.

SINGLE-CRYSTAL SILICON - Silicon grown from a single crystal which is sliced into thin wafers and used as the basic building block, the solar cell, in many PV module designs.

SOLAR ELECTRICITY - Electricity produced directly from solar energy.

STATE OF CHARGE - The percent of full charge that a battery currently has available.

SULFATION - The change in the normal crystalline structure of sulfate on the surface of a battery plate to more stable, larger crystals which increase resistance to normal battery charging.

VOLTAGE or VOLTS - The potential difference or "pressure" in an electrical circuit which results in the flow of electrons.

WATT - The unit of measure of electrical power.   The product of current and voltage: watts = amps x volts.

WATT HOUR - A measure of electrical energy equal to one watt of power used or produced over a one hour period of time.

Article 690 of the

National Electrical Code®

Article 690 of the *National Electrical Code®* is reprinted with permission from NFPA 70-1990, *National Electrical Code®*, Copyright© 1990, National Fire Protection Association, Quincy, MA 02269. This reprinted material is not the complete and official position of the NFPA, on the referenced subject which is represented only by the standard in its entirety.

ARTICLE 690 — SOLAR PHOTOVOLTAIC SYSTEMS

A. General

690-1. Scope. The provisions of this article apply to solar photovoltaic electrical energy systems including the array circuit(s), power conditioning unit(s) and controller(s) for such systems. Solar photovoltaic systems covered by this article may be interactive with other electric power production sources or stand alone, with or without electrical energy storage such as batteries. These systems may have alternating- or direct-current output for utilization.

690-2. Definitions.

Array: A mechanically integrated assembly of modules or panels with a support structure and foundation, tracking, thermal control, and other components, as required, to form a direct-current power-producing unit.

Blocking Diode: A diode used to block reverse flow of current into a photovoltaic source circuit.

Interactive System: A solar photovoltaic system that operates in parallel with and may be designed to deliver power to another electric power production source connected to the same load. For the purpose of this definition, an energy storage subsystem of a solar photovoltaic system, such as a battery, is not another electric power production source.

Module: The smallest complete, environmentally protected assembly of solar cells, optics and other components, exclusive of tracking, designed to generate direct-current power under sunlight.

Panel: A collection of modules mechanically fastened together, wired, and designed to provide a field-installable unit.

Photovoltaic Output Circuit: Circuit conductors between the photovoltaic source circuit(s) and the power conditioning unit or direct-current utilization equipment. See Diagram 690-1.

Photovoltaic Power Source: An array or aggregate of arrays which generates direct-current power at system voltage and current.

Photovoltaic Source Circuit: Conductors between modules and from modules to the common connection point(s) of the direct-current system. See Diagram 690-1.

Power Conditioning Unit: Equipment which is used to change voltage level or waveform or both of electrical energy. Commonly a power conditioning unit is an inverter which changes a direct-current input to an alternating-current output.

Power Conditioning Unit Output Circuit: Conductors between the power conditioning unit and the connection to the service equipment or another electric power production source such as a utility. See Diagram 690-1.

Solar Cell: The basic photovoltaic device which generates electricity when exposed to light.

Solar Photovoltaic System: The total components and subsystems which in combination convert solar energy into electrical energy suitable for connection to a utilization load.

Stand-Alone System: A solar photovoltaic system that supplies power independently but which may receive control power from another electric power production source.

690-3. Other Articles. Wherever the requirements of other articles of this Code and Article 690 differ, the requirements of Article 690 shall apply. Solar photovoltaic systems operating as interconnected power production sources shall be installed in accordance with the provisions of Article 705.

690-4. Installation.

(a) Photovoltaic System. A solar photovoltaic system shall be permitted to supply a building or other structure in addition to any service(s) of another electricity supply system(s).

(b) Conductors of Different Systems. Photovoltaic source circuits and photovoltaic output circuits shall not be contained in the same raceway, cable tray, cable, outlet box, junction box or similar fitting as feeders or branch circuits of other systems.

Exception: Where the conductors of the different systems are separated by a partition or are connected together.

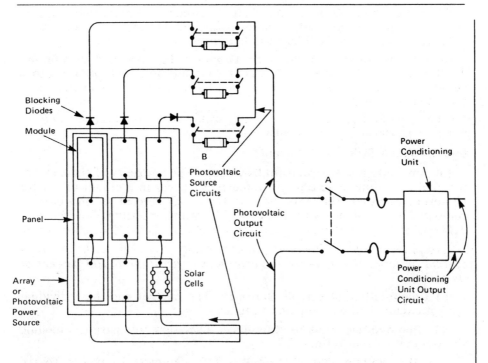

A: Disconnecting means required by Section 690-13.

B: Equipment permitted to be on the photovoltaic power source side of the photovoltaic power source disconnecting means, per Section 690-14, Exception No. 2. See Section 690-16.

Diagram 690-1.   Solar Photovoltaic System
(Simplified Circuit)

(c) Module Connection Arrangement.   The connections to a module or panel shall be so arranged that removal of a module or panel from a photovoltaic source circuit does not interrupt a grounded conductor to another photovoltaic source circuit.

690-5. Ground Fault Detection and Interruption.   Roof-mounted photovoltaic arrays located on dwellings shall be provided with ground-fault protection to reduce fire hazard. The ground-fault protection circuit shall be capable of detecting a ground fault, interrupting the fault path, and disabling the array.

B. Circuit Requirements

690-7. Maximum Voltage.

(a) Voltage Rating.   In a photovoltaic power source and its direct-current circuits, the voltage considered shall be the rated open-circuit voltage.

(b) Direct-Current Utilization Circuits.   The voltage of direct-current utilization circuits shall conform with Section 210-6.

(c) Photovoltaic Source and Output Circuits. Photovoltaic source circuits and photovoltaic output circuits which do not include lampholders, fixtures or receptacles shall be permitted up to 600 volts.

(d) Circuits Over 150 Volts to Ground. In one- and two-family dwellings, live parts in photovoltaic source circuits and photovoltaic output circuits over 150 volts to ground shall not be accessible while energized to other than qualified persons.

(FPN): See Section 110-17 for guarding of live parts, and Section 210-6 for voltage to ground and between conductors.

690-8. Circuit Sizing and Current.

(a) Ampacity and Overcurrent Devices. The ampacity of the conductors and the rating or setting of overcurrent devices in a circuit of a solar photovoltaic system shall not be less than 125 percent of the current computed in accordance with (b) below. The rating or setting of overcurrent devices shall be permitted in accordance with Section 240-3, Exception No. 4.

Exception: Circuits containing an assembly together with its overcurrent device(s) that is listed for continuous operation at 100 percent of its rating.

(b) Computation of Circuit Current. The current for the individual type of circuit shall be computed as follows:

(1) Photovoltaic Source Circuits. The sum of parallel module short-circuit current ratings.

(2) Photovoltaic Output Circuit. The photovoltaic power source current rating.

(3) Power Conditioning Unit Output Circuit. The power conditioning unit output current rating.

Exception: The current rating of a circuit without an overcurrent device shall be the short-circuit current, and it shall not exceed the ampacity of the circuit conductors.

690-9. Overcurrent Protection.

(a) Circuits and Equipment. Photovoltaic source circuit, photovoltaic output circuit, power conditioning unit output circuit, and storage battery circuit conductors and equipment shall be protected in accordance with the requirements of Article 240. Circuits connected to more than one electrical source shall have overcurrent devices so located as to provide overcurrent protection from all sources.

(FPN): Possible backfeed of current from any source of supply, including a supply through a power conditioning unit into the photovoltaic output circuit and photovoltaic source circuits, must be considered in determining whether adequate overcurrent protection from all sources is provided for conductors and modules.

(b) Power Transformers. Overcurrent protection for a transformer with a source(s) on each side shall be provided in accordance with Section 450-3 by considering first one side of the transformer, then the other side of the transformer as the primary.

Exception: A power transformer with a current rating on the side connected toward the photovoltaic power source not less than the short-circuit output current rating of the power conditioning unit shall be permitted without overcurrent protection from that source.

ARTICLE 690—SOLAR PHOTOVOLTAIC SYSTEMS   **70**-655

(c) Photovoltaic Source Circuits. Branch-circuit or supplementary type overcurrent devices shall be permitted to provide overcurrent protection in photovoltaic source circuits. The overcurrent devices shall be accessible, but shall not be required to be readily accessible.

C. Disconnecting Means

690-13. All Conductors. Means shall be provided to disconnect all current-carrying conductors of a photovoltaic power source from all other conductors in a building or other structure.

690-14. Additional Provisions. The provisions of Article 230, Part F shall apply to the photovoltaic power source disconnecting means.

Exception No. 1: The disconnecting means shall not be required to be suitable as service equipment and shall be rated in accordance with Section 690-17.

Exception No. 2: Equipment such as photovoltaic source circuit isolating switches, overcurrent devices, and blocking diodes shall be permitted on the photovoltaic power source side of the photovoltaic power source disconnecting means.

690-15. Disconnection of Photovoltaic Equipment. Means shall be provided to disconnect equipment, such as a power conditioning unit, filter assembly and the like from all ungrounded conductors of all sources. If the equipment is energized (live) from more than one source, the disconnecting means shall be grouped and identified.

690-16. Fuses. Disconnecting means shall be provided to disconnect a fuse from all sources of supply if the fuse is energized from both directions and is accessible to other than qualified persons. Such a fuse in a photovoltaic source circuit shall be capable of being disconnected independently of fuses in other photovoltaic source circuits.

690-17. Switch or Circuit Breaker. The disconnecting means for ungrounded conductors shall consist of a manually operable switch(es) or circuit breaker(s): (1) located where readily accessible, (2) externally operable without exposing the operator to contact with live parts, (3) plainly indicating whether in the open or closed position, and (4) having ratings not less than the load to be carried. Where all terminals of the disconnecting means may be energized in the open position, a warning sign shall be mounted on or adjacent to the disconnecting means. The sign shall be clearly legible and shall read substantially: WARNING - ELECTRIC SHOCK - DO NOT TOUCH - TERMINALS ENERGIZED IN OPEN POSITION.

Exception: A disconnecting means located on the direct-current side shall be permitted to have an interrupting rating less than the current-carrying rating when the system is designed so that the direct-current switch cannot be opened under load.

690-18. Disablement of an Array. Means shall be provided to disable an array or portions of an array.

(FPN): Photovoltaic modules are energized while exposed to light. Installation, replacement, or servicing of array components while a module(s) is irradiated may expose persons to electric shock.

70-656   NATIONAL ELECTRICAL CODE

D. Wiring Methods

690-31. Methods Permitted.

(a) Wiring Systems. All raceway and cable wiring methods included in this Code and other wiring systems and fittings specifically intended and identified for use on photovoltaic arrays shall be permitted. Where wiring devices with integral enclosures are used, sufficient length of cable shall be provided to facilitate replacement.

(b) Single Conductor Cable. Type UF single conductor cable shall be permitted in photovoltaic source circuits where installed in the same manner as a Type UF multiconductor cable in accordance with Article 339. Where exposed to direct rays of the sun, cable identified as sunlight-resistant shall be used.

690-32. Component Interconnections. Fittings and connectors which are intended to be concealed at the time of on-site assembly, when listed for such use, shall be permitted for on-site interconnection of modules or other array components. Such fittings and connectors shall be equal to the wiring method employed in insulation, temperature rise and fault-current withstand, and shall be capable of resisting the effects of the environment in which they are used.

690-33. Connectors. The connectors permitted by Section 690-32 shall comply with (a) through (e) below.

(a) Configuration. The connectors shall be polarized and shall have a configuration that is noninterchangeable with receptacles in other electrical systems on the premises.

(b) Guarding. The connectors shall be constructed and installed so as to guard against inadvertent contact with live parts by persons.

(c) Type. The connectors shall be of the latching or locking type.

(d) Grounding Member. The grounding member shall be the first to make and the last to break contact with the mating connector.

(e) Interruption of Circuit. The connectors shall be capable of interrupting the circuit current without hazard to the operator.

690-34. Access to Boxes. Junction, pull and outlet boxes located behind modules or panels shall be installed so that the wiring contained in them can be rendered accessible directly or by displacement of a module(s) or panel(s) secured by removable fasteners and connected by a flexible wiring system.

E. Grounding

690-41. System Grounding. For a photovoltaic power source, one conductor of a 2-wire system rated over 50 volts and a neutral conductor of a 3-wire system shall be solidly grounded.

Exception: Other methods which accomplish equivalent system protection and which utilize equipment listed and identified for the use shall be permitted.

(FPN): See the first Fine Print Note under Section 250-1.

690-42. Point of System Grounding Connection. The direct-current circuit grounding connection shall be made at any single point on the photovoltaic output circuit.

(FPN): Locating the grounding connection point as close as practicable to the photovoltaic source will better protect the system from voltage surges due to lightning.

690-43. Size of Equipment Grounding Conductor. The equipment grounding conductor shall be no smaller than the required size of the circuit conductors in systems where the available photovoltaic power source short-circuit current is less than twice the current rating of the overcurrent device. In other systems the equipment grounding conductor shall be sized in accordance with Section 250-95.

690-44. Common Grounding Electrode. Exposed noncurrent-carrying metal parts of equipment and conductor enclosures of a photovoltaic system shall be grounded to the grounding electrode that is used to ground the direct-current system. Two or more electrodes that are effectively bonded together shall be considered as a single electrode in this sense.

F. Marking

690-51. Modules. Modules shall be marked with identification of terminals or leads as to polarity, maximum overcurrent device rating for module protection and with rated: (1) open-circuit voltage, (2) operating voltage, (3) maximum permissible system voltage, (4) operating current, (5) short-circuit current, and (6) maximum power.

690-52. Photovoltaic Power Source. A marking, specifying the photovoltaic power source rated: (1) operating current, (2) operating voltage, (3) open-circuit voltage, and (4) short-circuit current, shall be provided at an accessible location at the disconnecting means for the photovoltaic power source.

(FPN): Reflecting systems used for irradiance enhancement may result in increased levels of output current and power.

G. Connection to Other Sources

690-61. Loss of System Voltage. The power output from a power conditioning unit in a solar photovoltaic system that is interactive with another electric system(s) shall be automatically disconnected from all ungrounded conductors in such other electric system(s) upon loss of voltage in that electric system(s) and shall not reconnect to that electric system(s) until its voltage is restored.

(FPN): For other interconnected electric power production sources, see Article 705.

A normally interactive solar photovoltaic system shall be permitted to operate as a stand-alone system to supply premises wiring.

690-62. Ampacity of Neutral Conductor. If a single-phase, 2-wire power conditioning unit output is connected to the neutral and one ungrounded conductor (only) of a 3-wire system or of a 3-phase, 4-wire wye-connected system, the maximum load connected between the neutral and any one ungrounded conductor plus the power conditioning unit output rating shall not exceed the ampacity of the neutral conductor.

690-63. Unbalanced Interconnections.

(a) Single-Phase. The output of a single-phase power conditioning unit shall not be connected to a 3-phase, 3- or 4-wire electrical service derived directly from a delta-connected transformer.

(b) Three-Phase. A 3-phase power conditioning unit shall be automatically disconnected from all ungrounded conductors of the interconnected system when one of the phases opens in either source.

Exception for (a) and (b): Where the interconnected system is designed so that significant unbalanced voltages will not result.

690-64. Point of Connection. The output of a power production source shall be connected as specified in (a) or (b) below.

(FPN): For the purposes of this section a power production source is considered to be: (1) the output of a power conditioning unit when connected to an alternating current electric source; and (2) the photovoltaic output circuit when interactive with a direct current electric source.

(a) Supply Side. To the supply side of the service disconnecting means as permitted in Section 230-82, Exception No. 6.

(b) Load Side. To the load side of the service disconnecting means of the other source(s), if all of the following conditions are met:

(1) Each source interconnection shall be made at a dedicated circuit breaker or fusible disconnecting means.

(2) The sum of the ampere ratings of overcurrent devices in circuits supplying power to a busbar or conductor shall not exceed the rating of the busbar or conductor.

Exception: For a dwelling unit the sum of the ampere ratings of the overcurrent devices shall not exceed 120 percent of the rating of the busbar or conductor.

(3) The interconnection point shall be on the line side of all ground-fault protection equipment.

Exception: Connection shall be permitted to be made to the load side of ground-fault protection provided that there is ground-fault protection for equipment from all ground-fault current sources.

(4) Equipment containing overcurrent devices in circuits supplying power to a busbar or conductor shall be marked to indicate the presence of all sources.

Exception: Equipment with power supplied from a single point of connection.

(5) Equipment such as circuit breakers, if back-fed, shall be identified for such operation.

H. Storage Batteries

690-71. Installation.

(a) General. Storage batteries in a solar photovoltaic system shall be installed in accordance with the provisions of Article 480.

Exception: As provided in Section 690-73.

(b) Dwellings.

(1) Storage batteries for dwellings shall have the cells connected so as to operate at less than 50 volts.

Exception: Where live parts are not accessible during routine battery maintenance, a battery system voltage in accordance with Section 690-7 shall be permitted.

(2) Live parts of battery systems for dwellings shall be guarded to prevent accidental contact by persons or objects, regardless of voltage or battery type.

(FPN): Batteries in solar photovoltaic systems are subject to extensive charge-discharge cycles and typically require frequent maintenance, such as checking electrolyte and cleaning connections.

690-72. State of Charge.   Equipment shall be provided to control the state of charge of the battery. All adjusting means for control of the state of charge shall be accessible only to qualified persons.

Exception: Where the design of the photovoltaic power source is matched to the voltage rating and charge current requirements for the interconnected battery cells.

690-73. Grounding.   The interconnected battery cells shall be considered grounded where the photovoltaic power source is installed in accordance with Section 690-41, Exception.

12 VOLT   2% WIRE LOSS TABLE

AMPS	WATTAGE	14ga	12ga	10ga	8ga	6ga	4ga	2ga	1/0	2/0	3/0
				DISTANCE in FEET							
1	12	45	70	115	180	290	456	720	~	~	~
2	24	22	35	57	90	145	228	360	580	720	912
4	48	10	17	27	45	72	114	180	290	360	456
6	72	7	12	17	30	47	75	120	193	243	305
8	96	5	8	14	22	35	57	90	145	180	228
10	120	4	7	11	18	28	45	72	115	145	183
15	180	3	4	7	12	19	30	48	76	96	122
20	240	#	3	5	9	14	22	36	57	72	91
25	300	#	#	4	7	11	18	29	46	58	73
30	360	#	#	3	6	9	15	24	38	48	61
40	480	#	#	#	4	7	11	18	29	36	45
50	600	#	#	#	#	5	9	14	23	29	36

12 VOLT   5% WIRE LOSS TABLE

AMPS	WATTAGE	14ga	12ga	10ga	8ga	6ga	4ga	2ga	1/0	2/0	3/0
				DISTANCE in FEET							
1	12	113	175	275	450	710	~	~	~	~	~
2	24	56	87	138	225	355	576	900	~	~	~
4	48	25	43	68	113	178	288	450	725	900	~
6	72	18	30	43	75	119	188	300	481	600	760
8	96	13	21	36	56	88	144	225	363	450	570
10	120	11	17	28	45	71	113	180	290	360	457
15	180	7	11	17	30	47	75	120	193	240	304
20	240	#	8	13	22	36	56	90	145	180	229
25	300	#	#	11	17	28	45	72	115	145	183
30	360	#	#	8	15	23	37	60	96	120	152
40	480	#	#	#	11	17	28	45	72	90	114
50	600	#	#	#	#	13	22	36	57	72	91

12 VOLT   10% WIRE LOSS TABLE

AMPS	WATTAGE	14ga	12ga	10ga	8ga	6ga	4ga	2ga	1/0	2/0	3/0
				DISTANCE in FEET							
1	12	225	350	575	900	~	~	~	~	~	~
2	24	113	175	288	450	726	~	~	~	~	~
4	48	50	87	138	225	363	563	908	~	~	~
6	72	37	60	87	150	238	375	600	963	~	~
8	96	27	42	72	113	178	285	450	725	~	~
10	120	22	35	57	90	143	228	363	575	725	915
15	180	15	22	35	60	95	150	240	383	480	610
20	240	#	17	27	45	77	113	180	288	363	458
25	300	#	#	22	35	57	90	145	230	290	365
30	360	#	#	17	30	47	75	120	193	243	305
40	480	#	#	#	22	35	57	90	145	180	228
50	600	#	#	#	#	27	45	72	115	145	183

~   OVER 1000 FEET
#   EXCEEDS AMPACITY
NOTE: DISTANCE   IS FROM POWER SOURCE TO THE LOAD NOT   THE   ROUND TRIP DISTANCE.

24 VOLT   2% WIRE LOSS TABLE

AMPS	WATTAGE	14ga	12ga	10ga	8ga	6ga	4ga	2ga	1/0	2/0	3/0
				DISTANCE in FEET				~	~	~	~
1	24	90	140	230	360	580	912	~	~	~	~
2	48	45	70	115	180	290	456	720	~	~	~
4	96	20	35	55	90	145	228	360	580	720	912
6	144	15	24	35	60	95	150	240	386	486	610
8	192	11	17	29	45	71	114	180	290	360	456
10	240	9	14	23	36	57	91	145	230	290	366
15	360	6	9	14	24	38	60	96	153	192	244
20	480	#	7	11	18	29	45	72	115	145	183
25	600	#	#	9	14	23	36	58	92	116	146
30	720	#	#	7	12	19	30	48	77	97	122
40	960	#	#	#	9	14	23	36	58	72	91
50	1200	#	#	#	#	11	18	29	46	58	73

24 VOLT   5% WIRE LOSS TABLE

AMPS	WATTAGE	14ga	12ga	10ga	8ga	6ga	4ga	2ga	1/0	2/0	3/0
				DISTANCE in FEET			~	~	~	~	~
1	24	226	350	550	900	~	~	~	~	~	~
2	48	112	175	276	450	710	~	~	~	~	~
4	96	50	87	137	226	356	576	900	~	~	~
6	144	37	60	87	150	238	376	600	962	~	~
8	192	27	42	72	112	177	288	450	726	900	~
10	240	22	35	57	90	142	226	360	580	720	914
15	360	15	22	35	60	95	150	240	386	480	608
20	480	#	17	27	45	72	112	180	290	360	458
25	600	#	#	22	35	57	90	145	230	290	366
30	720	#	#	17	30	47	75	120	192	240	304
40	960	#	#	#	23	35	57	90	145	180	228
50	1200	#	#	#	#	27	45	72	115	145	182

24 VOLT   10% WIRE LOSS TABLE

AMPS	WATTAGE	14ga	12ga	10ga	8ga	6ga	4ga	2ga	1/0	2/0	3/0
				DISTANCE in FEET	~	~	~	~	~	~	~
1	24	450	700	~	~	~	~	~	~	~	~
2	48	226	350	576	900	~	~	~	~	~	~
4	96	100	175	276	450	726	~	~	~	~	~
6	144	75	120	175	300	476	750	~	~	~	~
8	192	55	85	145	226	356	570	900	~	~	~
10	240	45	70	115	180	286	456	726	~	~	~
15	360	30	45	70	120	190	300	480	766	960	~
20	480	#	35	55	90	145	226	360	576	726	916
25	600	#	#	45	70	115	180	290	460	580	730
30	720	#	#	35	60	95	150	240	386	486	610
40	960	#	#	#	45	70	115	180	290	360	456
50	1200	#	#	#	#	55	90	145	230	290	366

~ OVER 1000 FEET
# EXCEEDS AMPACITY
NOTE: DISTANCE IS FROM POWER SOURCE TO THE LOAD NOT THE ROUND
TRIP DISTANCE.

120 VOLT 2% WIRE LOSS TABLE

AMPS	WATTAGE	14ga	12ga	10ga	8ga	6ga	4ga	2ga	1/0	2/0	3/0
				DISTANCE in FEET							
1	120	450	700	~	~	~	~	~	~	~	~
2	240	225	350	575	900	~	~	~	~	~	~
4	480	100	175	275	450	725	~	~	~	~	~
6	720	75	120	175	275	450	725	~	~	~	~
8	960	55	85	145	225	355	570	~	~	~	~
10	1200	45	70	120	190	300	480	765	960	~	~
15	1800	30	45	70	120	190	300	480	765	960	~
20	2400	#	35	55	90	145	225	360	575	725	915
25	3000	#	#	45	70	115	180	290	460	580	730
30	3600	#	#	#	60	95	150	240	385	485	610
40	4800	#	#	#	45	70	115	180	290	360	455
50	6000	#	#	#	#	55	90	145	230	290	365

120 VOLT 5% WIRE LOSS TABLE

AMPS	WATTAGE	14ga	12ga	10ga	8ga	6ga	4ga	2ga	1/0	2/0	3/0
				DISTANCE in FEET							
1	120	~	~	~	~	~	~	~	~	~	~
2	240	563	875	~	~	~	~	~	~	~	~
4	480	250	438	688	~	~	~	~	~	~	~
6	720	188	300	438	750	~	~	~	~	~	~
8	960	138	213	363	563	888	~	~	~	~	~
10	1200	113	175	288	450	713	~	~	~	~	~
15	1800	75	113	175	300	475	750	~	~	~	~
20	2400	#	87	138	225	363	563	900	~	~	~
25	3000	#	#	113	175	288	450	725	~	~	~
30	3600	#	#	87	150	238	375	600	963	~	~
40	4800	#	#	#	113	175	288	450	725	900	~
50	6000	#	#	#	#	138	228	363	575	725	913

120 VOLT 10% WIRE LOSS TABLE

AMPS	WATTAGE	14ga	12ga	10ga	8ga	6ga	4ga	2ga	1/0	2/0	3/0
				DISTANCE in FEET							
1	120	~	~	~	~	~	~	~	~	~	~
2	240	~	~	~	~	~	~	~	~	~	~
4	480	500	876	~	~	~	~	~	~	~	~
6	720	376	600	876	~	~	~	~	~	~	~
8	960	276	426	726	~	~	~	~	~	~	~
10	1200	226	350	576	900	~	~	~	~	~	~
15	1800	150	226	350	600	950	~	~	~	~	~
20	2400	100	175	276	450	726	~	~	~	~	~
25	3000	#	#	226	350	576	900	~	~	~	~
30	3600	#	#	175	300	476	750	~	~	~	~
40	4800	#	#	#	226	350	576	900	~	~	~
50	6000	#	#	#	#	276	456	726	~	~	~

~ OVER 1000 FEET
# EXCEEDS AMPACITY
NOTE: DISTANCE IS FROM POWER SOURCE TO THE LOAD NOT THE ROUND TRIP DISTANCE.

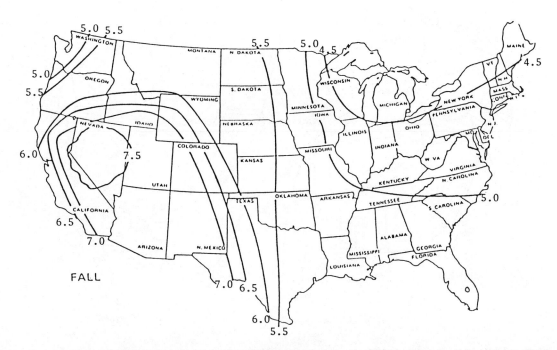

AVERAGE DAILY PEAK SUN HOURS FOR FALL SEASON (Aug., Sept., Oct.) FOR ARRAY ANGLE EQUAL TO LATITUDE (Courtesy of DOE)

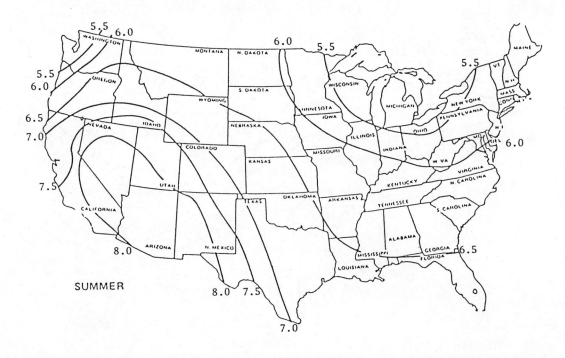

AVERAGE DAILY PEAK SUN HOURS FOR SUMMER SEASON (May, June, July) FOR ARRAY ANGLE OF LAT. (-) 15 Deg. (Courtesy of DOE)

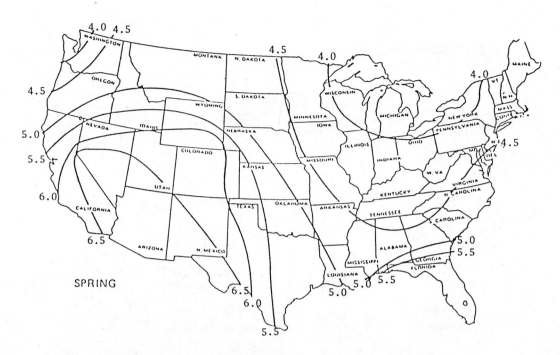

AVERAGE DAILY PEAK SUN HOURS FOR SPRING SEASON (Feb., Mar., Apr.) FOR ARRAY ANGLE EQUAL TO LATITUDE (Courtesy of DOE)

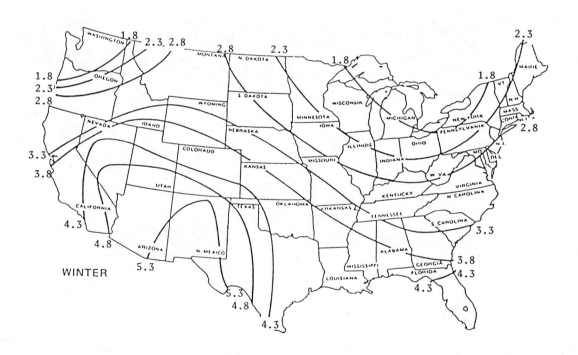

AVERAGE DAILY PEAK SUN HOURS FOR WINTER SEASON (Nov., Dec., Jan.) FOR ARRAY ANGLE OF LAT. (+) 15 Deg. (Courtesy of DOE)

INSOLATION DATA BY SEASON (Courtesy of DOE)

AVERAGE DAILY INSOLATION AVAILABILITY
FOR A SURFACE TILTED AT LATITUDE - 15 DEGREES (KWH/M2)

SITE	WINTER D,J,F	SPRING M,A,M	SUMMER J,J,A	FALL S,O,N	ANNUAL
ALBUQUERQUE	4.92	7.59	7.90	6.35	6.70
ATLANTA	3.12	5.51	5.85	4.43	4.74
AUSTIN	3.78	5.35	6.42	4.84	5.10
BIRMINGHAM	3.08	5.35	5.58	4.47	4.62
BISMARCK	2.96	5.36	6.73	4.07	4.79
BOSTON	2.36	4.49	5.40	3.59	3.97
BROWNSVILLE	3.49	5.72	6.67	4.74	5.16
BRYCE CANYON	4.87	7.39	7.65	6.24	6.55
CARIBOU	2.46	5.09	5.35	2.97	3.98
COLUMBIA	2.90	5.39	6.34	4.38	4.77
DAGGETT	4.63	7.59	8.23	6.14	6.66
DALLAS-FORT WORTH	3.53	5.43	6.77	4.95	5.18
DENVER	4.52	6.75	7.16	5.75	6.06
DETROIT	2.13	4.65	5.72	3.57	4.03
ELKO	4.12	6.62	8.15	5.85	6.19
EL PASO	5.11	7.72	7.83	6.24	6.73
FAIRBANKS	0.54	5.03	4.83	1.96	3.11
FRESNO	3.17	7.12	8.33	5.90	6.14
GREAT FALLS	2.69	5.47	7.04	4.27	4.88
HONOLULU	4.19	5.76	6.27	5.11	5.34
LAS VEGAS	5.02	8.03	8.19	6.35	6.91
MADISON	2.68	5.04	5.95	3.75	4.36
MEDFORD	1.98	5.55	7.60	4.25	4.87
MIAMI	4.13	5.72	5.33	4.57	4.94
NASHVILLE	2.61	5.01	6.02	4.08	4.44
NEW ORLEANS	3.42	5.55	5.65	4.68	4.84
OKLAHOMA CITY	3.67	5.66	6.63	4.98	5.24
OMAHA	3.46	5.46	6.53	4.38	4.97
ORLANDO	4.02	6.00	5.51	4.68	5.06
PHOENIX	4.77	7.80	7.84	6.26	6.68
PITTSBURGH	1.90	4.33	5.39	3.41	3.76
RALEIGH-DURHAM	3.00	5.22	5.66	4.16	4.52
SACRAMENTO	3.07	6.76	8.32	5.62	5.96
SAN DIEGO	4.31	6.21	6.67	5.25	5.62
SAN JUAN	4.51	5.83	5.92	4.99	5.32
SEATTLE	1.37	4.46	5.92	2.89	3.67
SYRACUSE	1.71	4.29	5.41	3.02	3.62
WASHINGTON D.C.	2.80	4.95	5.60	3.89	4.32

AVERAGE DAILY INSOLATION AVAILABILITY
FOR A SOUTH FACING SURFACE TILTED AT LATITUDE DEGREES (KWH/M2)

SITE	WINTER	SPRING	SUMMER	FALL	ANNUAL
ALBUQUERQUE	5.70	7.57	7.49	6.92	6.92
ATLANTA	3.53	5.48	5.57	4.78	4.85
AUSTIN	4.32	5.34	6.08	5.21	5.24
BIRMINGHAM	3.47	5.32	5.32	4.80	4.73
BISMARCK	3.36	5.35	6.39	4.35	4.87
BOSTON	2.67	4.46	5.14	3.84	4.03
BROWNSVILLE	3.92	5.68	6.30	5.05	5.25
BRYCE CANYON	5.62	7.34	7.16	6.76	6.73
CARIBOU	2.77	5.09	5.10	3.15	4.04
COLUMBIA	3.28	5.37	6.01	4.71	4.85
DAGGETT	5.32	7.52	7.71	6.64	6.81
DALLAS-FORT WORTH	4.03	5.43	6.41	5.33	5.31
DENVER	5.22	6.73	6.76	6.22	6.24
DETROIT	2.39	4.64	5.43	3.78	4.07
ELKO	4.71	6.58	7.66	6.29	6.32
EL PASO	5.91	7.68	7.39	6.79	6.95
FAIRBANKS	0.60	5.02	4.62	2.05	3.09
FRESNO	3.54	7.05	7.82	6.34	6.20
GREAT FALLS	3.04	5.47	6.66	4.57	4.95
HONOLULU	4.74	5.73	5.93	5.49	5.48
LAS VEGAS	5.79	7.96	7.67	6.88	7.08
MADISON	3.02	5.05	5.66	3.99	4.14
MEDFORD	2.19	5.53	7.18	4.51	4.87
MIAMI	4.69	5.72	5.10	4.92	5.11
NASHVILLE	2.94	5.00	5.75	4.36	4.52
NEW ORLEANS	3.86	5.52	5.38	5.04	4.96
OKLAHOMA CITY	4.18	5.65	6.28	5.37	5.38
OMAHA	3.93	5.43	6.20	4.71	5.08
ORLANDO	4.57	5.96	5.26	5.05	5.21
PHOENIX	5.49	7.77	7.42	6.80	6.98
PITTSBURGH	2.11	4.29	5.15	3.62	3.80
RALEIGH-DURHAM	3.37	5.21	5.39	4.46	4.61
SACRAMENTO	3.47	6.70	7.80	6.03	6.01
SAN DIEGO	4.94	6.20	6.35	5.67	5.79
SAN JUAN	5.14	5.80	5.61	5.36	5.48
SEATTLE	1.51	4.43	5.66	3.03	3.67
SYRACUSE	1.90	4.27	5.16	3.20	3.64
WASHINGTON D.C.	3.17	4.93	5.33	4.16	4.41

INSOLATION DATA BY SEASONS (Courtesy of DOE)

AVERAGE DAILY INSOLATION AVAILABILITY
FOR A SOUTH FACING SURFACE TILTED AT LATITUDE + 15 DEGREES (KWH/M2)

SITE	WINTER D,J,F	SPRING M,A,M	SUMMER J,J,A	FALL S,O,N	ANNUAL
ALBUQUERQUE	6.14	7.18	6.72	7.12	6.79
ATLANTA	3.78	5.20	5.03	4.91	4.73
AUSTIN	4.63	5.11	5.45	5.33	5.13
BIRMINGHAM	3.70	5.05	4.84	4.91	4.63
BISMARCK	3.57	5.09	5.73	4.43	4.71
BOSTON	2.83	4.24	4.66	3.89	3.91
BROWNSVILLE	4.18	5.42	5.64	5.15	5.10
BRYCE CANYON	6.04	6.91	6.32	6.90	6.55
CARIBOU	2.92	4.88	4.62	3.20	3.91
COLUMBIA	3.49	5.10	5.41	4.80	4.71
DAGGETT	5.70	7.07	6.80	6.77	6.59
DALLAS-FORT WORTH	4.31	5.21	5.75	5.45	5.18
DENVER	5.60	6.37	6.01	6.35	6.09
DETROIT	2.53	4.42	4.91	3.82	3.93
ELKO	5.03	6.22	6.78	6.38	6.11
EL PASO	6.38	7.28	6.60	6.99	6.81
FAIRBANKS	0.62	4.77	4.23	2.07	2.93
FRESNO	3.76	6.64	6.90	6.44	5.95
GREAT FALLS	3.23	5.22	5.95	4.64	4.77
HONOLULU	5.07	5.45	5.34	5.60	5.37
LAS VEGAS	6.21	7.49	6.78	7.02	6.88
MADISON	3.20	4.84	5.12	4.04	4.31
MEDFORD	2.30	5.25	6.40	4.54	4.64
MIAMI	5.01	5.47	4.67	5.07	5.05
NASHVILLE	3.12	4.77	5.24	4.44	4.40
NEW ORLEANS	4.11	5.26	4.88	5.16	4.85
OKLAHOMA CITY	4.48	5.39	5.64	5.49	5.25
OMAHA	4.19	5.16	5.58	4.80	4.94
ORLANDO	4.88	5.66	4.79	5.19	5.13
PHOENIX	5.91	7.36	6.65	6.99	6.73
PITTSBURGH	2.23	4.09	4.68	3.66	3.68
RALEIGH-DURHAM	3.57	4.97	4.88	4.55	4.49
SACRAMENTO	3.68	6.30	6.89	6.12	5.76
SAN DIEGO	5.29	5.90	5.73	5.80	5.68
SAN JUAN	5.51	5.53	5.04	5.48	5.39
SEATTLE	1.59	4.24	5.13	3.05	3.52
SYRACUSE	2.00	4.08	4.70	3.24	3.51
WASHINGTON D.C.	3.37	4.70	4.84	4.24	4.29

AVERAGE DAILY INSOLATION AVAILABILITY
FOR A SURFACE TRACKING ABOUT A NORTH - SOUTH AXIS
TILTED AT LATITUDE - 15 DEGREES (KWH/M2)

SITE	WINTER	SPRING	SUMMER	FALL	ANNUAL
ALBUQUERQUE	6.34	10.13	10.60	8.32	8.86
ATLANTA	3.89	7.03	7.41	5.51	5.97
AUSTIN	4.80	6.75	8.27	6.14	6.50
BIRMINGHAM	3.80	6.78	7.00	5.57	5.80
BISMARCK	3.57	7.03	9.17	5.14	6.25
BOSTON	2.81	5.72	7.03	4.51	5.03
BROWNSVILLE	4.32	7.08	8.70	5.97	6.53
BRYCE CANYON	6.20	10.28	10.78	8.24	8.89
CARIBOU	2.93	6.61	7.14	3.63	5.09
COLUMBIA	3.44	6.84	8.55	5.55	6.12
DAGGETT	5.74	10.15	11.27	7.88	8.78
DALLAS-FORT WORTH	4.39	6.95	8.93	6.35	6.67
DENVER	5.76	9.09	9.73	7.49	8.03
DETROIT	2.54	5.88	7.36	4.33	5.04
ELKO	5.10	9.05	11.43	7.60	8.32
EL PASO	6.63	10.35	10.43	8.25	8.93
FAIRBANKS	0.60	6.67	6.85	2.33	4.14
FRESNO	3.80	9.55	11.55	7.57	8.15
GREAT FALLS	3.20	7.07	9.70	5.44	6.37
HONOLULU	5.21	7.12	7.91	6.48	6.69
LAS VEGAS	6.33	10.99	11.18	8.24	9.21
MADISON	3.17	6.49	7.67	4.62	5.50
MEDFORD	2.29	7.09	10.36	5.30	6.28
MIAMI	5.24	7.22	6.47	5.76	6.19
NASHVILLE	3.17	6.16	7.52	5.07	5.49
NEW ORLEANS	4.18	7.02	7.09	5.86	6.05
OKLAHOMA CITY	4.61	7.35	8.62	6.32	6.74
OMAHA	4.27	7.15	8.84	5.59	6.48
ORLANDO	4.99	7.66	6.82	5.82	6.33
PHOENIX	6.04	10.47	10.33	8.05	8.74
PITTSBURGH	2.23	5.37	6.80	4.12	4.55
RALEIGH-DURHAM	3.56	6.63	7.16	5.03	5.61
SACRAMENTO	3.67	8.96	11.46	7.11	7.82
SAN DIEGO	5.38	7.84	8.38	6.48	7.03
SAN JUAN	5.73	7.17	7.40	6.31	6.66
SEATTLE	1.57	5.61	7.82	3.49	4.64
SYRACUSE	2.00	5.41	6.88	3.63	4.50
WASHINGTON D.C.	3.40	6.33	7.20	4.80	5.45

INSOLATION DATA BY SEASONS (Courtesy of DOE)

AVERAGE DAILY INSOLATION AVAILABILITY
FOR A SURFACE TRACKING ABOUT A NORTH - SOUTH AXIS
TILTED AT LATITUDE DEGREES (KWH/M2)

SITE	WINTER D.J.F	SPRING M.A.M	SUMMER J.J.A	FALL S.O.N	ANNUAL
ALBUQUERQUE	6.94	10.13	10.35	8.73	9.05
ATLANTA	4.21	7.03	7.26	5.77	6.08
AUSTIN	5.22	6.76	8.08	6.43	6.63
BIRMINGHAM	4.11	6.78	6.85	5.82	5.90
BISMARCK	3.89	7.05	9.00	5.37	6.14
BOSTON	3.07	5.72	6.89	4.69	5.10
BROWNSVILLE	4.67	7.08	8.48	6.21	6.62
BRYCE CANYON	6.80	10.29	10.53	8.64	9.08
CARIBOU	3.18	6.63	7.01	3.78	5.17
COLUMBIA	3.77	6.85	8.38	5.80	6.21
DAGGETT	6.30	10.15	11.00	8.26	8.95
DALLAS-FORT WORTH	4.79	6.97	8.73	6.65	6.80
DENVER	6.31	9.10	9.51	7.86	8.21
DETROIT	2.76	5.88	7.21	4.49	5.10
ELKO	5.58	9.08	11.18	7.95	8.46
EL PASO	7.24	10.35	10.17	8.65	9.12
FAIRBANKS	0.64	6.70	6.75	2.41	4.15
FRESNO	4.12	9.55	11.28	7.91	8.24
GREAT FALLS	3.50	7.09	9.50	5.68	6.46
HONOLULU	5.67	7.11	7.72	6.77	6.82
LAS VEGAS	6.96	10.99	10.91	8.64	9.39
MADISON	3.46	6.52	7.52	4.81	5.59
MEDFORD	2.47	7.10	10.14	5.49	6.32
MIAMI	5.69	7.23	6.34	6.03	6.32
NASHVILLE	3.43	6.17	7.37	5.29	5.58
NEW ORLEANS	4.53	7.02	6.93	6.13	6.17
OKLAHOMA CITY	5.02	7.36	8.43	6.63	6.87
OMAHA	4.64	7.16	8.66	5.85	6.59
ORLANDO	5.43	7.66	6.66	6.12	6.47
PHOENIX	6.61	10.46	10.08	8.45	8.91
PITTSBURGH	2.41	5.37	6.67	4.29	4.70
RALEIGH-DURHAM	3.86	6.64	7.01	5.26	5.70
SACRAMENTO	3.99	8.96	11.20	7.43	7.91
SAN DIEGO	5.89	7.86	8.20	6.81	7.19
SAN JUAN	6.23	7.16	7.22	6.59	6.81
SEATTLE	1.69	5.61	7.67	3.60	4.56
SYRACUSE	2.16	5.41	6.75	3.77	4.53
WASHINGTON D.C.	3.70	6.34	7.05	5.02	5.54

AVERAGE DAILY INSOLATION AVAILABILITY
FOR A SURFACE TRACKING ABOUT A NORTH - SOUTH AXIS
TILTED AT LATITUDE + 15 DEGREES (KWH/M2)

SITE	WINTER	SPRING	SUMMER	FALL	ANNUAL
ALBUQUERQUE	7.24	9.85	9.78	8.85	8.94
ATLANTA	4.38	6.85	6.90	5.86	6.01
AUSTIN	5.44	6.61	7.64	6.51	6.56
BIRMINGHAM	4.28	6.61	6.52	5.90	5.83
BISMARCK	4.04	6.88	8.55	5.42	6.24
BOSTON	3.19	5.58	6.55	4.73	5.02
BROWNSVILLE	4.86	6.89	8.03	6.27	6.52
BRYCE CANYON	7.10	10.03	10.00	8.74	8.98
CARIBOU	3.30	6.50	6.71	3.80	5.09
COLUMBIA	3.92	6.66	7.99	5.86	6.12
DAGGETT	6.58	9.86	10.40	8.36	8.81
DALLAS-FORT WORTH	5.00	6.80	8.28	6.73	6.72
DENVER	6.59	8.87	9.02	7.95	8.12
DETROIT	2.86	5.74	6.86	4.53	5.01
ELKO	5.81	8.85	10.63	8.02	8.34
EL PASO	7.56	10.05	9.60	8.76	9.00
FAIRBANKS	0.66	6.53	6.51	2.42	4.05
FRESNO	4.28	9.29	10.70	7.99	8.08
GREAT FALLS	3.63	6.92	9.02	5.73	6.34
HONOLULU	5.89	6.91	7.29	6.86	6.74
LAS VEGAS	7.26	10.70	10.33	8.75	9.27
MADISON	3.58	6.37	7.15	4.86	5.50
MEDFORD	2.54	6.91	9.62	5.52	6.17
MIAMI	5.91	7.07	6.03	6.12	6.28
NASHVILLE	3.57	6.01	7.01	5.34	5.49
NEW ORLEANS	4.72	6.84	6.59	6.22	6.10
OKLAHOMA CITY	5.23	7.20	8.00	6.70	6.79
OMAHA	4.83	7.00	8.26	5.91	6.51
ORLANDO	5.66	7.47	6.34	6.21	6.42
PHOENIX	6.89	10.16	9.51	8.56	8.79
PITTSBURGH	2.50	5.24	6.35	4.32	4.61
RALEIGH-DURHAM	4.01	6.48	6.65	5.33	5.63
SACRAMENTO	4.14	8.71	10.61	7.49	7.76
SAN DIEGO	6.13	7.66	7.76	6.90	7.12
SAN JUAN	6.50	6.98	6.82	6.68	6.75
SEATTLE	1.74	5.47	7.32	3.63	4.56
SYRACUSE	2.23	5.28	6.42	3.80	4.45
WASHINGTON D.C.	3.84	6.18	6.71	5.08	5.46

MAP OF DEVIATION OF MAGNETIC NORTH FROM TRUE NORTH
(Adapted from Isogonic Chart of the USA, courtesy of Dept. of Commerce)

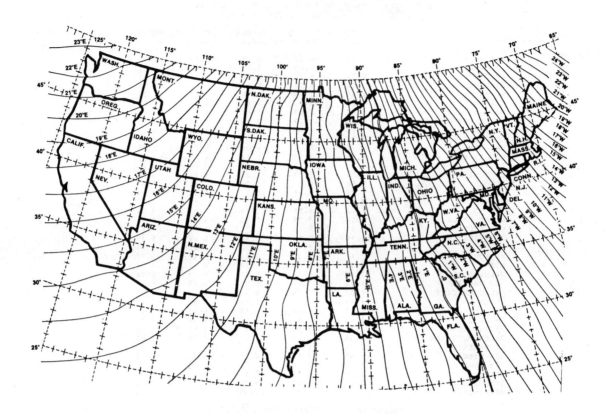

APPLIANCES AND WATTAGE RATINGS

AC HOUSEHOLD APPLIANCES	AVERAGE WATTAGE
BLENDER	290 - 385
CEILING FAN	120
COMPUTER	100
DISHWASHER	1190 - 1250
FLUORESCENT BULB, GE COMPAX	15
FREEZER	340 - 600
HAIR DRIER	500 - 1500
IRON	1000
MICROWAVE OVEN	300 - 1450
RADIO	50
REFRIGERATOR	330
SEWING MACHINE	100
TELEVISION, COLOR (recent)	100
TOASTER	1100 - 1250
VACUUM CLEANER	600
VCR	40
WASHING MACHINE	375 - 550

AC SHOP TOOLS and PUMPS	AVERAGE WATTAGE
1/3 HP DEEP WELL PUMP	900
1/2 HP DEEP WELL PUMP	1200
CIRCULATING PUMP	80
BELT SANDER	600
DRILL, 3/8 in	240
SAW, CIRCULAR	800 - 1200
SAW, TABLE	800 - 950

DC APPLIANCES	AVERAGE WATTAGE
BLENDER	80
BOOSTER PUMP (FLOWLIGHT)	240
CAR STEREO, 12 VT	10
CEILING FAN	25
FAN, 8 in	12 - 30
FLUORESCENT LIGHT, PL TUBE	13
FLUORESCENT LIGHT, THINLIGHT	15
PHONE MACHINE, 12 VT	2
RADIO/TAPE PLAYER	35
REFRIGERATOR	60
TELEVISION, BLACK & WHITE	18
TELEVISION, COLOR	60

PART A: LOAD ESTIMATES

DC APPLIANCE   HRS/DAY   X   WATTS   =   WATT-HRS/DAY

_____

_____

_____

_____

_____

_____

_____

_____

_____

_____

                    DC AVERAGE DAILY LOADS _____

AC APPLIANCE   HRS/DAY   X   WATTS   =   WATT-HRS/DAY

_____

_____

_____

_____

_____

_____

_____

_____

_____

            subtotal AC AVERAGE DAILY LOADS_____
     [15% of above] 15% INVERTER ALLOWANCE_____
              TOTAL AC AVERAGE DAILY LOADS_____

                    DC AVERAGE DAILY LOADS_____

     TOTAL [AC AND DC] AVERAGE DAILY LOADS_____

PART B: PV SYSTEM SIZING WORKSHEET

SECTION 1 - SITE DATA AND MODULE SPECIFICATIONS
LOCATION _____LATITUDE_____
MODULE MODEL_____PEAK WATTS_____ NOMINAL VOLTAGE_____
AMPS at typical load_____

SECTION 2 - ARRAY ANGLE ADJUSTMENTS
LATITUDE (-) 15 degrees = SUMMER ARRAY ANGLE   A_____
LATITUDE   = FALL   ARRAY ANGLE   B_____
LATITUDE (+) 15 degrees = WINTER ARRAY ANGLE   C_____
LATITUDE   = SPRING ARRAY ANGLE   D_____
LATITUDE (+) 15 degrees = FIXED ARRAY ANGLE   C_____

SECTION 3 - USABLE OUTPUT OF MODULE
MODULE AMPS at typical load E_____X NOMINAL MODULE VOLTAGE____
= USABLE MODULE OUTPUT in watts F_____

SECTION 4 - AVERAGE DAILY PEAK SUN HOURS from maps in appendix
SUMMER AVERAGE DAILY PEAK SUN HOURS at lat.(-)15 deg. G_____
FALL   AVERAGE DAILY PEAK SUN HOURS at lat.   H_____
WINTER AVERAGE DAILY PEAK SUN HOURS at lat.(+)15 deg. I_____
SPRING AVERAGE DAILY PEAK SUN HOURS at lat.   J_____
FIXED AVERAGE DAILY PEAK SUN HOURS = (FALL A.D.P.S.Hrs.
H_____ + SPRING A.D.P.S. Hrs. J_____ / 2 = K_____

SECTION 5 - AVERAGE DAILY USABLE MODULE OUTPUT
SUMMER A.D.P.S. Hrs. G_____ X USABLE MOD. OUTPUT F_____
= SUMMER AVER. DAILY USABLE MOD. OUTPUT L_____

FALL   A.D.P.S. Hrs. H_____ X USABLE MOD. OUTPUT F_____
= FALL   AVER. DAILY USABLE MOD. OUTPUT M_____

WINTER A.D.P.S. Hrs. I_____ X USABLE MOD. OUTPUT F_____
= WINTER AVER. DAILY USABLE MOD. OUTPUT N_____

SPRING A.D.P.S. Hrs. J_____ X USABLE MOD. OUTPUT F_____
= SPRING AVER. DAILY USABLE MOD. OUPUT O_____

FIXED A.D.P.S. Hrs. K_____ X USABLE MOD. OUTPUT F_____
= FIXED AVER. DAILY USABLE MOD. OUTPUT P_____

SECTION 6 - ARRAY SIZING
AVERAGE DAILY LOADS_____ / USABLE MODULE OUTPUT_____
= NUMBER OF MODULES Q_____

SECTION 7 - BATTERY BANK SIZING
BATTERY VOLTAGE_____ X AMP-HRS at 20 hr. rate_____= BATTERY
SIZE in watt-hrs R_____

30 WATT-HRS X MODULE PEAK WATTS_____X NUMBER OF MODULES Q_____
= BATTERY BANK WATT-HRS S_____

BATTERY BANK S_____ / BATTERY SIZE R_____= NUMBER OF
BATTERIES T_____

SECTION 7 - ESTIMATED ARRAY USABLE OUTPUT
Here you choose a number of modules and a season to get a budget
for loads.
PROSPECTIVE NUMBER OF MODULES U_____
SEASONAL ARRAY ANGLE from Sect.4 SEASON_____ANGLE_____
AVERAGE DAILY MODULE OUTPUT from Sect.5
NUMBER OF MODULES U____ X AVER. DAILY MOD. OUTPUT_____ = AVER.
DAILY ARRAY OUTPUT V_____

AVERAGE DAILY ARRAY OUTPUT V_____ will be the average
number of watt-hrs available to power your loads for the season
and the angle you have chosen for your prospective system. You
may repeat the calculations for another season. Use Section 7 for
sizing battery bank for prospective system.

INSOLATION DATA BY MONTH

SITE ARRAY TILT		JAN	FEB	MAR	APR	MAY	JUN	JUL	AUG	SEP	OCT	NOV	DEC	ANNUAL TOTAL (KWH/SQ. M)	AVERAGE DAY (KWH/SQ. M)
JUNEAU	AK	LATITUDE:	58 DEGREES	22 MINUTES											
LATITUDE -15:		1.04	1.70	2.57	3.72	4.07	4.28	4.02	3.47	2.73	1.74	1.24	.68	951.8	2.6
LATITUDE :		1.16	1.81	2.57	3.53	3.71	3.86	3.67	3.28	2.72	1.83	1.39	.78	922.7	2.5
LATITUDE +15:		1.22	1.82	2.43	3.18	3.24	3.32	3.20	2.96	2.58	1.83	1.45	.83	853.6	2.3
KING SALMON	AK	LATITUDE:	58 DEGREES	41 MINUTES											
LATITUDE -15:		1.25	2.22	3.68	4.40	4.73	4.68	4.38	3.73	3.54	3.08	2.07	1.08	1183.5	3.2
LATITUDE :		1.39	2.36	3.72	4.19	4.32	4.22	4.00	3.54	3.57	3.32	2.35	1.25	1164.2	3.2
LATITUDE +15:		1.46	2.37	3.56	3.78	3.77	3.63	3.48	3.19	3.41	3.36	2.49	1.34	1090.5	3.0
KODIAK	AK	LATITUDE:	57 DEGREES	45 MINUTES											
LATITUDE -15:		1.13	1.92	3.50	4.38	4.35	4.65	4.46	4.20	3.55	3.06	1.90	.99	1160.0	3.2
LATITUDE :		1.26	2.03	3.53	4.16	3.98	4.20	4.08	3.99	3.58	3.29	2.15	1.13	1137.8	3.1
LATITUDE +15:		1.31	2.02	3.37	3.76	3.47	3.61	3.55	3.60	3.42	3.32	2.27	1.20	1062.9	2.9
NOME	AK	LATITUDE:	64 DEGREES	30 MINUTES											
LATITUDE -15:		.34	1.56	3.29	4.63	5.14	5.40	4.57	3.74	3.53	2.59	1.16	.01	1095.5	3.0
LATITUDE :		.38	1.65	3.32	4.39	4.69	4.85	4.15	3.54	3.58	2.79	1.32	.01	1055.8	2.9
LATITUDE +15:		.40	1.65	3.17	3.96	4.06	4.14	3.60	3.19	3.42	2.82	1.39	.01	968.4	2.7
BIRMINGHAM	AL	LATITUDE:	33 DEGREES	34 MINUTES											
LATITUDE -15:		2.75	3.53	4.44	5.39	5.72	5.83	5.58	5.57	5.01	4.54	3.44	2.70	1658.7	4.5
LATITUDE :		3.02	3.73	4.49	5.19	5.31	5.33	5.17	5.36	5.08	4.87	3.84	3.05	1657.0	4.5
LATITUDE +15:		3.14	3.74	4.31	4.75	4.68	4.63	4.57	4.91	4.89	4.94	4.03	3.23	1577.2	4.3
MOBILE	AL	LATITUDE:	30 DEGREES	41 MINUTES											
LATITUDE -15:		3.11	3.92	4.74	5.51	5.78	5.69	5.30	5.25	4.89	4.72	3.64	2.94	1688.3	4.6
LATITUDE :		3.43	4.16	4.80	5.31	5.37	5.22	4.93	5.06	4.95	5.06	4.07	3.32	1694.8	4.6
LATITUDE +15:		3.58	4.19	4.63	4.87	4.75	4.55	4.36	4.64	4.77	5.14	4.28	3.53	1620.9	4.4
MONTGOMERY	AL	LATITUDE:	32 DEGREES	18 MINUTES											
LATITUDE -15:		2.87	3.66	4.56	5.56	5.85	5.99	5.67	5.62	5.01	4.67	3.59	2.87	1703.1	4.7
LATITUDE :		3.17	3.88	4.61	5.36	5.43	5.47	5.26	5.42	5.08	5.01	4.02	3.25	1703.0	4.7
LATITUDE +15:		3.29	3.89	4.43	4.91	4.78	4.75	4.64	4.96	4.89	5.08	4.22	3.45	1622.1	4.4
FORT SMITH	AR	LATITUDE:	35 DEGREES	20 MINUTES											
LATITUDE -15:		3.03	3.75	4.56	5.23	5.90	6.33	6.37	6.11	5.27	4.64	3.55	2.94	1756.2	4.8
LATITUDE :		3.37	3.98	4.61	5.03	5.46	5.76	5.88	5.89	5.35	4.98	3.99	3.34	1754.8	4.8
LATITUDE +15:		3.52	4.00	4.43	4.60	4.81	4.98	5.16	5.38	5.15	5.05	4.20	3.56	1669.3	4.6
LITTLE ROCK	AR	LATITUDE:	34 DEGREES	44 MINUTES											
LATITUDE -15:		2.93	3.73	4.54	5.21	5.95	6.38	6.27	6.04	5.30	4.71	3.47	2.84	1746.4	4.8
LATITUDE :		3.24	3.96	4.59	5.01	5.51	5.82	5.80	5.82	5.38	5.06	3.89	3.22	1744.7	4.8
LATITUDE +15:		3.39	3.98	4.41	4.59	4.85	5.03	5.09	5.32	5.19	5.14	4.09	3.43	1659.4	4.5
PHOENIX	AZ	LATITUDE:	33 DEGREES	26 MINUTES											
LATITUDE -15:		4.19	5.21	6.36	7.67	8.26	8.28	7.66	7.46	7.13	6.11	4.83	4.02	2348.9	6.4
LATITUDE :		4.74	5.62	6.51	7.40	7.60	7.48	7.06	7.20	7.32	6.66	5.52	4.66	2366.3	6.5
LATITUDE +15:		5.02	5.72	6.31	6.76	6.60	6.36	6.15	6.57	7.10	6.83	5.89	5.03	2260.9	6.2
PRESCOTT	AZ	LATITUDE:	34 DEGREES	39 MINUTES											
LATITUDE -15:		4.30	5.14	6.30	7.44	8.11	8.34	7.12	6.82	6.97	6.09	4.94	4.15	2304.2	6.3
LATITUDE :		4.88	5.54	6.45	7.18	7.47	7.53	6.57	6.58	7.16	6.64	5.66	4.83	2327.1	6.4
LATITUDE +15:		5.18	5.64	6.24	6.56	6.50	6.41	5.74	6.01	6.94	6.81	6.05	5.23	2229.6	6.1
TUCSON	AZ	LATITUDE:	32 DEGREES	7 MINUTES											
LATITUDE -15:		4.42	5.35	6.47	7.66	8.23	8.24	7.21	7.06	6.90	6.08	4.94	4.18	2335.1	6.4
LATITUDE :		5.00	5.77	6.63	7.39	7.57	7.44	6.65	6.81	7.07	6.62	5.65	4.85	2356.1	6.5
LATITUDE +15:		5.30	5.87	6.42	6.74	6.57	6.33	5.81	6.22	6.84	6.79	6.02	5.25	2255.1	6.2
WINSLOW	AZ	LATITUDE:	35 DEGREES	1 MINUTES											
LATITUDE -15:		4.19	5.15	6.34	7.47	8.01	8.19	7.24	7.00	6.90	6.01	4.90	4.04	2295.2	6.3
LATITUDE :		4.75	5.55	6.49	7.21	7.37	7.39	6.67	6.74	7.07	6.54	5.61	4.69	2314.0	6.3
LATITUDE +15:		5.03	5.64	6.28	6.57	6.40	6.29	5.81	6.15	6.84	6.70	5.99	5.07	2213.3	6.1

Stand-Alone Flat-plate Photovoltaic Power Systems: System Sizing and Life-Cycle Costing Methodology for Federal Agencies (Courtesy of DOE)

SITE ARRAY TILT		JAN	FEB	MAR	APR	MAY	JUN	JUL	AUG	SEP	OCT	NOV	DEC	ANNUAL TOTAL (KWH/SQ. M)	AVERAGE DAY (KWH/SQ. M)
DAGGETT	CA	LATITUDE:		34 DEGREES	52 MINUTES										
LATITUDE -15:		4.02	4.92	6.29	7.44	8.01	8.36	8.03	7.82	7.20	5.98	4.68	3.90	2334.1	6.4
LATITUDE :		4.55	5.30	6.44	7.18	7.38	7.55	7.40	7.56	7.41	6.53	5.35	4.52	2348.6	6.4
LATITUDE +15:		4.82	5.38	6.24	6.56	6.42	6.43	6.43	6.90	7.19	6.69	5.71	4.89	2241.3	6.1
EL TORO	CA	LATITUDE:		33 DEGREES	40 MINUTES										
LATITUDE -15:		3.85	4.65	5.60	6.25	6.39	6.65	7.29	7.00	6.07	5.17	4.24	3.73	2036.4	5.6
LATITUDE :		4.33	4.98	5.71	6.02	5.91	6.05	6.73	6.76	6.20	5.59	4.81	4.30	2051.4	5.6
LATITUDE +15:		4.57	5.05	5.52	5.51	5.20	5.22	5.87	6.17	5.99	5.69	5.10	4.63	1963.7	5.4
FRESNO	CA	LATITUDE:		36 DEGREES	46 MINUTES										
LATITUDE -15:		2.72	3.89	5.59	6.89	7.69	8.26	8.31	8.04	7.27	5.81	3.90	2.50	2158.6	5.9
LATITUDE :		3.01	4.14	5.70	6.64	7.08	7.46	7.64	7.76	7.48	6.33	4.41	2.83	2146.5	5.9
LATITUDE +15:		3.14	4.17	5.51	6.07	6.18	6.36	6.63	7.08	7.26	6.48	4.68	3.00	2025.7	5.5
LONG BEACH	CA	LATITUDE:		33 DEGREES	49 MINUTES										
LATITUDE -15:		3.77	4.56	5.61	6.28	6.37	6.48	7.09	6.83	5.94	5.05	4.14	3.62	2001.8	5.5
LATITUDE :		4.24	4.89	5.72	6.06	5.90	5.91	6.55	6.59	6.06	5.45	4.69	4.18	2015.9	5.5
LATITUDE +15:		4.48	4.95	5.53	5.54	5.18	5.11	5.73	6.02	5.86	5.55	4.98	4.49	1929.4	5.3
LOS ANGELES	CA	LATITUDE:		33 DEGREES	56 MINUTES										
LATITUDE -15:		3.78	4.57	5.65	6.32	6.36	6.43	7.12	6.76	5.88	5.02	4.16	3.65	1999.6	5.5
LATITUDE :		4.25	4.89	5.76	6.10	5.89	5.86	6.57	6.52	5.99	5.41	4.72	4.21	2014.3	5.5
LATITUDE +15:		4.48	4.96	5.57	5.58	5.18	5.07	5.75	5.96	5.79	5.51	5.00	4.53	1928.6	5.3
NEEDLES	CA	LATITUDE:		34 DEGREES	46 MINUTES										
LATITUDE -15:		4.15	5.23	6.49	7.59	8.19	8.44	7.84	7.45	7.21	6.07	4.87	4.09	2362.5	6.5
LATITUDE :		4.70	5.65	6.65	7.33	7.54	7.62	7.22	7.19	7.42	6.62	5.58	4.76	2381.9	6.5
LATITUDE +15:		4.98	5.76	6.45	6.69	6.56	6.48	6.29	6.57	7.20	6.79	5.96	5.16	2277.5	6.2
OAKLAND	CA	LATITUDE:		37 DEGREES	44 MINUTES										
LATITUDE -15:		3.05	3.97	5.22	6.32	6.85	7.12	7.19	6.78	6.19	4.89	3.66	2.98	1955.4	5.4
LATITUDE :		3.40	4.23	5.31	6.09	6.32	6.46	6.62	6.53	6.33	5.28	4.13	3.40	1951.7	5.3
LATITUDE +15:		3.56	4.27	5.12	5.56	5.54	5.55	5.78	5.96	6.12	5.37	4.37	3.64	1851.7	5.1
RED BLUFF	CA	LATITUDE:		40 DEGREES	9 MINUTES										
LATITUDE -15:		2.53	3.58	4.95	6.36	7.39	7.87	8.31	7.79	7.00	5.24	3.29	2.44	2033.9	5.6
LATITUDE :		2.80	3.79	5.02	6.11	6.80	7.11	7.63	7.51	7.19	5.68	3.70	2.77	2013.7	5.5
LATITUDE +15:		2.92	3.81	4.83	5.57	5.93	6.07	6.61	6.85	6.96	5.79	3.89	2.94	1893.3	5.2
SACRAMENTO	CA	LATITUDE:		38 DEGREES	31 MINUTES										
LATITUDE -15:		2.55	3.69	5.27	6.63	7.55	8.11	8.35	7.92	7.11	5.47	3.53	2.45	2090.9	5.7
LATITUDE :		2.81	3.91	5.36	6.38	6.95	7.33	7.67	7.65	7.31	5.94	3.98	2.77	2073.2	5.7
LATITUDE +15:		2.93	3.94	5.17	5.83	6.06	6.25	6.65	6.97	7.09	6.07	4.20	2.94	1951.6	5.3
SAN DIEGO	CA	LATITUDE:		32 DEGREES	44 MINUTES										
LATITUDE -15:		3.90	4.70	5.64	6.25	6.18	6.26	6.74	6.65	5.95	5.16	4.31	3.79	1994.8	5.5
LATITUDE :		4.39	5.04	5.75	6.03	5.72	5.72	6.23	6.42	6.07	5.58	4.89	4.37	2014.9	5.5
LATITUDE +15:		4.63	5.11	5.56	5.52	5.04	4.95	5.46	5.87	5.87	5.68	5.19	4.71	1934.5	5.3
SAN FRANCISCO	CA	LATITUDE:		37 DEGREES	37 MINUTES										
LATITUDE -15:		3.04	3.93	5.22	6.31	6.90	7.19	7.41	7.00	6.35	4.94	3.64	2.98	1975.5	5.4
LATITUDE :		3.39	4.19	5.30	6.08	6.36	6.52	6.82	6.75	6.49	5.33	4.11	3.36	1970.1	5.4
LATITUDE +15:		3.55	4.22	5.11	5.55	5.57	5.60	5.95	6.16	6.28	5.43	4.34	3.59	1867.3	5.1
SANTA MARIA	CA	LATITUDE:		34 DEGREES	54 MINUTES										
LATITUDE -15:		3.51	4.31	5.56	6.25	6.60	7.12	7.22	6.88	6.11	5.26	4.11	3.52	2023.8	5.5
LATITUDE :		3.93	4.61	5.66	6.03	6.11	6.46	6.67	6.64	6.25	5.70	4.66	4.06	2032.8	5.6
LATITUDE +15:		4.14	4.66	5.47	5.51	5.36	5.56	5.82	6.06	6.04	5.82	4.95	4.36	1940.4	5.3
SUNNYVALE	CA	LATITUDE:		37 DEGREES	25 MINUTES										
LATITUDE -15:		3.17	4.04	5.32	6.39	7.05	7.42	7.55	7.17	6.41	5.03	3.74	3.02	2019.1	5.5
LATITUDE :		3.54	4.31	5.41	6.15	6.50	6.72	6.95	6.91	6.55	5.43	4.22	3.45	2013.9	5.5
LATITUDE +15:		3.72	4.34	5.21	5.62	5.68	5.76	6.05	6.30	6.34	5.53	4.46	3.69	1908.7	5.2
COLORADO SPRINGS	CO	LATITUDE:		38 DEGREES	49 MINUTES										
LATITUDE -15:		4.17	4.81	5.66	6.39	6.60	7.17	6.86	6.72	6.51	5.72	4.50	3.92	2100.9	5.8
LATITUDE :		4.74	5.18	5.77	6.15	6.09	6.50	6.32	6.48	6.67	6.23	5.16	4.57	2126.1	5.8
LATITUDE +15:		5.04	5.27	5.57	5.62	5.34	5.59	5.53	5.91	6.45	6.38	5.50	4.95	2042.8	5.6
DENVER	CO	LATITUDE:		39 DEGREES	45 MINUTES										
LATITUDE -15:		4.01	4.65	5.65	6.24	6.62	7.12	7.05	6.82	6.45	5.55	4.29	3.75	2076.2	5.7
LATITUDE :		4.56	5.01	5.76	6.00	6.11	6.46	6.50	6.57	6.60	6.04	4.90	4.36	2095.9	5.7
LATITUDE +15:		4.84	5.08	5.56	5.48	5.35	5.55	5.67	6.00	6.39	6.17	5.22	4.72	2009.0	5.5
GRAND JUNCTION	CO	LATITUDE:		39 DEGREES	7 MINUTES										
LATITUDE -15:		3.65	4.56	5.70	6.59	7.38	7.86	7.65	7.28	6.86	5.70	4.41	3.66	2171.0	5.9
LATITUDE :		4.12	4.89	5.81	6.34	6.79	7.10	7.03	7.02	7.03	6.20	5.03	4.25	2180.1	6.0
LATITUDE +15:		4.35	4.96	5.61	5.78	5.92	6.06	6.12	6.39	6.81	6.34	5.36	4.58	2078.0	5.7

SITE ARRAY TILT	JAN	FEB	MAR	APR	MAY	JUN	JUL	AUG	SEP	OCT	NOV	DEC	ANNUAL TOTAL (KWH/SQ. M)	AVERAGE DAY (KWH/SQ. M)
PUEBLO CO LATITUDE: 38 DEGREES 17 MINUTES														
LATITUDE -15:	4.12	4.73	5.68	6.46	6.69	7.36	7.15	6.97	6.55	5.67	4.49	3.86	2122.9	5.8
LATITUDE :	4.68	5.09	5.79	6.22	6.17	6.67	6.59	6.72	6.71	6.16	5.12	4.48	2142.6	5.9
LATITUDE +15:	4.97	5.16	5.59	5.68	5.40	5.71	5.74	6.13	6.49	6.30	5.46	4.85	2053.1	5.6
HARTFORD CT LATITUDE: 41 DEGREES 56 MINUTES														
LATITUDE -15:	2.14	2.85	3.51	4.33	4.86	5.11	5.11	4.70	4.22	3.53	2.27	1.81	1352.9	3.7
LATITUDE :	2.35	3.00	3.53	4.15	4.49	4.66	4.72	4.51	4.26	3.76	2.51	2.02	1337.8	3.7
LATITUDE +15:	2.44	2.99	3.37	3.78	3.96	4.05	4.16	4.11	4.08	3.79	2.61	2.13	1262.1	3.5
GUANTANAMO BAY CU LATITUDE: 19 DEGREES 54 MINUTES														
LATITUDE -15:	4.66	5.38	6.17	6.69	6.36	6.10	6.50	6.31	5.84	5.17	4.76	4.39	2079.2	5.7
LATITUDE :	5.27	5.79	6.31	6.48	5.93	5.62	6.06	6.12	5.96	5.56	5.40	5.07	2115.9	5.8
LATITUDE +15:	5.60	5.90	6.12	5.95	5.23	4.89	5.33	5.62	5.77	5.67	5.74	5.47	2046.1	5.6
WASHINGTON-STERLINDC LATITUDE: 38 DEGREES 57 MINUTES														
LATITUDE -15:	2.44	3.15	3.98	4.77	5.32	5.76	5.62	5.30	4.81	4.02	2.86	2.16	1528.7	4.2
LATITUDE :	2.69	3.32	4.01	4.58	4.93	5.26	5.20	5.10	4.87	4.29	3.19	2.43	1518.5	4.2
LATITUDE +15:	2.80	3.33	3.85	4.19	4.35	4.56	4.58	4.66	4.68	4.34	3.34	2.57	1437.8	3.9
WILMINGTON DE LATITUDE: 39 DEGREES 40 MINUTES														
LATITUDE -15:	2.49	3.25	4.11	4.85	5.30	5.71	5.65	5.32	4.76	3.97	2.90	2.26	1539.3	4.2
LATITUDE :	2.76	3.43	4.14	4.65	4.90	5.21	5.22	5.11	4.82	4.25	3.23	2.55	1530.0	4.2
LATITUDE +15:	2.87	3.44	3.97	4.25	4.32	4.51	4.59	4.66	4.63	4.29	3.39	2.70	1449.3	4.0
APALACHICOLA FL LATITUDE: 29 DEGREES 44 MINUTES														
LATITUDE -15:	3.16	3.98	4.94	6.02	6.45	6.08	5.60	5.39	5.16	4.93	3.94	3.12	1788.9	4.9
LATITUDE :	3.49	4.22	5.01	5.81	5.98	5.56	5.21	5.21	5.24	5.31	4.42	3.54	1795.8	4.9
LATITUDE +15:	3.65	4.25	4.84	5.32	5.26	4.83	4.61	4.78	5.05	5.40	4.67	3.77	1716.4	4.7
DAYTONA BEACH FL LATITUDE: 29 DEGREES 11 MINUTES														
LATITUDE -15:	3.56	4.29	5.19	6.02	6.07	5.57	5.51	5.36	4.94	4.45	3.89	3.32	1769.5	4.8
LATITUDE :	3.96	4.56	5.28	5.81	5.63	5.11	5.12	5.18	5.00	4.75	4.35	3.77	1780.2	4.9
LATITUDE +15:	4.15	4.60	5.09	5.32	4.96	4.46	4.53	4.75	4.81	4.80	4.58	4.02	1705.4	4.7
JACKSONVILLE FL LATITUDE: 30 DEGREES 30 MINUTES														
LATITUDE -15:	3.40	4.16	5.14	5.95	6.03	5.73	5.56	5.42	4.85	4.40	3.81	3.18	1753.7	4.8
LATITUDE :	3.77	4.42	5.22	5.74	5.60	5.25	5.17	5.23	4.91	4.71	4.27	3.61	1761.6	4.8
LATITUDE +15:	3.95	4.46	5.03	5.26	4.94	4.57	4.57	4.80	4.73	4.76	4.50	3.85	1685.2	4.6
MIAMI FL LATITUDE: 25 DEGREES 48 MINUTES														
LATITUDE -15:	3.74	4.50	5.27	5.90	5.71	5.25	5.46	5.17	4.76	4.46	3.99	3.69	1761.3	4.8
LATITUDE :	4.17	4.80	5.36	5.70	5.33	4.84	5.09	5.00	4.83	4.76	4.47	4.21	1781.2	4.9
LATITUDE +15:	4.38	4.85	5.18	5.24	4.72	4.24	4.52	4.60	4.65	4.82	4.71	4.50	1715.6	4.7
ORLANDO FL LATITUDE: 28 DEGREES 33 MINUTES														
LATITUDE -15:	3.68	4.37	5.29	6.06	6.14	5.60	5.56	5.33	4.99	4.61	4.08	3.50	1801.8	4.9
LATITUDE :	4.10	4.65	5.38	5.86	5.71	5.14	5.18	5.15	5.06	4.94	4.59	3.99	1817.2	5.0
LATITUDE +15:	4.31	4.70	5.19	5.37	5.03	4.49	4.58	4.73	4.87	5.01	4.85	4.26	1744.9	4.8
TALLAHASSEE FL LATITUDE: 30 DEGREES 23 MINUTES														
LATITUDE -15:	3.30	4.05	4.98	5.84	5.98	5.73	5.39	5.36	5.03	4.78	3.86	3.16	1748.3	4.8
LATITUDE :	3.66	4.31	5.06	5.63	5.55	5.25	5.02	5.17	5.10	5.13	4.33	3.58	1757.6	4.8
LATITUDE +15:	3.82	4.34	4.87	5.16	4.89	4.57	4.44	4.74	4.91	5.21	4.56	3.81	1682.6	4.6
TAMPA FL LATITUDE: 27 DEGREES 58 MINUTES														
LATITUDE -15:	3.68	4.39	5.30	6.09	6.18	5.65	5.42	5.26	4.95	4.73	4.08	3.49	1802.0	4.9
LATITUDE :	4.10	4.68	5.40	5.88	5.74	5.19	5.05	5.09	5.02	5.07	4.59	3.98	1819.3	5.0
LATITUDE +15:	4.31	4.73	5.22	5.40	5.07	4.53	4.48	4.67	4.84	5.15	4.85	4.25	1748.7	4.8
ATLANTA GA LATITUDE: 33 DEGREES 39 MINUTES														
LATITUDE -15:	2.80	3.55	4.47	5.43	5.72	5.81	5.59	5.51	4.89	4.50	3.55	2.76	1661.6	4.6
LATITUDE :	3.08	3.75	4.52	5.23	5.31	5.31	5.19	5.31	4.96	4.82	3.98	3.12	1661.3	4.6
LATITUDE +15:	3.21	3.76	4.34	4.79	4.68	4.62	4.58	4.86	4.77	4.89	4.18	3.32	1582.5	4.3
AUGUSTA GA LATITUDE: 33 DEGREES 22 MINUTES														
LATITUDE -15:	2.94	3.72	4.59	5.58	5.75	5.78	5.55	5.37	4.84	4.57	3.69	2.97	1684.7	4.6
LATITUDE :	3.25	3.94	4.64	5.37	5.34	5.29	5.15	5.17	4.89	4.90	4.14	3.37	1687.5	4.6
LATITUDE +15:	3.38	3.96	4.46	4.92	4.71	4.59	4.55	4.74	4.71	4.97	4.36	3.58	1610.3	4.4
MACON GA LATITUDE: 32 DEGREES 42 MINUTES														
LATITUDE -15:	2.97	3.70	4.65	5.60	5.81	5.83	5.50	5.53	4.92	4.63	3.74	2.94	1699.3	4.7
LATITUDE :	3.28	3.92	4.71	5.39	5.39	5.34	5.11	5.33	4.99	4.97	4.20	3.34	1703.3	4.7
LATITUDE +15:	3.42	3.94	4.53	4.94	4.76	4.64	4.51	4.88	4.80	5.05	4.42	3.55	1626.3	4.5
SAVANNAH GA LATITUDE: 32 DEGREES 8 MINUTES														
LATITUDE -15:	3.06	3.77	4.75	5.66	5.72	5.60	5.49	5.21	4.63	4.48	3.70	3.02	1677.2	4.6
LATITUDE :	3.38	3.99	4.81	5.46	5.31	5.13	5.10	5.02	4.68	4.79	4.15	3.43	1681.1	4.6
LATITUDE +15:	3.52	4.01	4.63	5.00	4.68	4.47	4.50	4.59	4.49	4.85	4.36	3.65	1605.4	4.4

SITE ARRAY TILT		JAN	FEB	MAR	APR	MAY	JUN	JUL	AUG	SEP	OCT	NOV	DEC	ANNUAL TOTAL (KWH/SQ. M)	AVERAGE DAY (KWH/SQ. M)
BARBERS POINT	HI	LATITUDE:		21 DEGREES	19 MINUTES										
LATITUDE -15:		4.06	4.75	5.29	5.79	6.13	6.26	6.28	6.22	5.85	5.15	4.39	3.97	1951.9	5.3
LATITUDE :		4.54	5.07	5.38	5.61	5.72	5.74	5.85	6.02	5.96	5.53	4.94	4.54	1974.8	5.4
LATITUDE +15:		4.78	5.14	5.21	5.16	5.05	4.98	5.15	5.53	5.77	5.63	5.23	4.86	1900.3	5.2
HILO	HI	LATITUDE:		19 DEGREES	43 MINUTES										
LATITUDE -15:		3.69	4.04	4.29	4.52	4.85	5.17	5.07	5.02	4.94	4.46	3.64	3.37	1614.9	4.4
LATITUDE :		4.09	4.29	4.35	4.39	4.56	4.79	4.76	4.88	5.01	4.76	4.03	3.81	1634.2	4.5
LATITUDE +15:		4.29	4.32	4.21	4.06	4.08	4.21	4.25	4.50	4.85	4.81	4.23	4.04	1577.0	4.3
HONOLULU	HI	LATITUDE:		21 DEGREES	20 MINUTES										
LATITUDE -15:		3.97	4.60	5.22	5.67	6.07	6.21	6.22	6.21	5.84	5.10	4.27	3.85	1924.4	5.3
LATITUDE :		4.43	4.90	5.31	5.49	5.66	5.70	5.80	6.01	5.95	5.48	4.80	4.39	1944.7	5.3
LATITUDE +15:		4.65	4.96	5.13	5.05	5.00	4.95	5.11	5.52	5.76	5.57	5.07	4.69	1869.8	5.1
BURLINGTON	IA	LATITUDE:		40 DEGREES	47 MINUTES										
LATITUDE -15:		2.63	3.46	4.21	5.07	5.82	6.42	6.48	6.09	5.23	4.45	3.11	2.31	1684.2	4.6
LATITUDE :		2.92	3.67	4.26	4.86	5.38	5.84	5.98	5.86	5.32	4.79	3.49	2.62	1674.1	4.6
LATITUDE +15:		3.06	3.68	4.08	4.44	4.73	5.03	5.23	5.35	5.12	4.86	3.68	2.78	1584.1	4.3
DES MOINES	IA	LATITUDE:		41 DEGREES	32 MINUTES										
LATITUDE -15:		2.71	3.52	4.31	5.16	5.80	6.43	6.51	6.12	5.36	4.57	3.16	2.42	1707.7	4.7
LATITUDE :		3.02	3.74	4.35	4.94	5.35	5.84	6.00	5.88	5.45	4.92	3.55	2.75	1699.3	4.7
LATITUDE +15:		3.16	3.76	4.18	4.51	4.70	5.04	5.25	5.36	5.25	5.00	3.74	2.93	1609.3	4.4
MASON CITY	IA	LATITUDE:		43 DEGREES	9 MINUTES										
LATITUDE -15:		2.71	3.53	4.34	5.06	5.90	6.40	6.49	6.19	5.34	4.46	3.00	2.31	1697.1	4.6
LATITUDE :		3.02	3.75	4.39	4.84	5.44	5.80	5.97	5.95	5.43	4.80	3.37	2.62	1685.9	4.6
LATITUDE +15:		3.17	3.77	4.21	4.41	4.76	5.00	5.22	5.42	5.23	4.86	3.54	2.79	1593.7	4.4
SIOUX CITY	IA	LATITUDE:		42 DEGREES	24 MINUTES										
LATITUDE -15:		2.73	3.49	4.31	5.24	5.91	6.44	6.60	6.20	5.36	4.51	3.17	2.40	1716.6	4.7
LATITUDE :		3.04	3.71	4.35	5.03	5.45	5.84	6.08	5.96	5.45	4.86	3.57	2.73	1707.0	4.7
LATITUDE +15:		3.18	3.72	4.17	4.58	4.78	5.03	5.31	5.43	5.25	4.93	3.76	2.90	1615.2	4.4
BOISE	ID	LATITUDE:		43 DEGREES	34 MINUTES										
LATITUDE -15:		2.33	3.57	4.95	6.18	7.13	7.48	8.20	7.55	6.87	5.20	3.23	2.30	1980.0	5.4
LATITUDE :		2.58	3.80	5.03	5.94	6.56	6.76	7.53	7.28	7.06	5.65	3.64	2.62	1962.5	5.4
LATITUDE +15:		2.69	3.82	4.84	5.41	5.73	5.79	6.53	6.63	6.84	5.77	3.84	2.78	1847.2	5.1
LEWISTON	ID	LATITUDE:		46 DEGREES	23 MINUTES										
LATITUDE -15:		1.62	2.59	3.88	4.84	5.77	6.11	7.36	6.70	5.76	3.98	2.13	1.50	1593.0	4.4
LATITUDE :		1.76	2.71	3.91	4.63	5.31	5.54	6.76	6.44	5.87	4.27	2.37	1.68	1562.4	4.3
LATITUDE +15:		1.81	2.70	3.74	4.21	4.65	4.77	5.88	5.86	5.66	4.33	2.47	1.76	1458.1	4.0
POCATELLO	ID	LATITUDE:		42 DEGREES	55 MINUTES										
LATITUDE -15:		2.59	3.73	5.19	6.14	7.13	7.53	8.15	7.67	6.95	5.47	3.54	2.49	2028.8	5.6
LATITUDE :		2.89	3.97	5.28	5.90	6.57	6.81	7.49	7.40	7.14	5.95	4.01	2.85	2018.7	5.5
LATITUDE +15:		3.02	4.00	5.09	5.38	5.74	5.84	6.51	6.75	6.93	6.09	4.25	3.04	1907.2	5.2
CHICAGO	IL	LATITUDE:		41 DEGREES	47 MINUTES										
LATITUDE -15:		2.30	3.05	4.03	4.82	5.55	6.09	6.04	5.74	5.03	4.09	2.65	1.91	1562.9	4.3
LATITUDE :		2.54	3.22	4.06	4.62	5.13	5.54	5.57	5.51	5.11	4.39	2.96	2.14	1546.6	4.2
LATITUDE +15:		2.64	3.22	3.89	4.22	4.51	4.78	4.88	5.03	4.91	4.44	3.10	2.26	1457.9	4.0
SPRINGFIELD	IL	LATITUDE:		39 DEGREES	50 MINUTES										
LATITUDE -15:		2.58	3.41	4.08	4.97	5.78	6.35	6.38	5.98	5.33	4.40	3.09	2.28	1663.6	4.6
LATITUDE :		2.86	3.61	4.12	4.77	5.34	5.78	5.88	5.75	5.42	4.73	3.46	2.58	1653.3	4.5
LATITUDE +15:		2.98	3.63	3.95	4.36	4.70	4.99	5.15	5.25	5.22	4.80	3.63	2.74	1564.5	4.3
FORT WAYNE	IN	LATITUDE:		41 DEGREES	0 MINUTES										
LATITUDE -15:		1.95	2.73	3.50	4.46	5.18	5.57	5.53	5.27	4.65	3.80	2.31	1.65	1419.2	3.9
LATITUDE :		2.13	2.85	3.51	4.27	4.78	5.07	5.10	5.05	4.70	4.05	2.54	1.83	1397.2	3.8
LATITUDE +15:		2.19	2.84	3.35	3.89	4.21	4.40	4.48	4.60	4.51	4.09	2.64	1.91	1311.9	3.6
INDIANAPOLIS	IN	LATITUDE:		39 DEGREES	44 MINUTES										
LATITUDE -15:		2.09	2.88	3.67	4.58	5.23	5.66	5.59	5.42	4.79	3.95	2.55	1.84	1469.6	4.0
LATITUDE :		2.28	3.02	3.69	4.39	4.84	5.17	5.17	5.20	4.85	4.22	2.82	2.05	1452.7	4.0
LATITUDE +15:		2.36	3.02	3.53	4.01	4.27	4.48	4.54	4.75	4.66	4.26	2.95	2.15	1369.2	3.8
SOUTH BEND	IN	LATITUDE:		41 DEGREES	42 MINUTES										
LATITUDE -15:		1.77	2.57	3.56	4.57	5.34	5.83	5.75	5.55	4.76	3.78	2.25	1.52	1439.3	3.9
LATITUDE :		1.92	2.69	3.57	4.38	4.93	5.30	5.30	5.33	4.82	4.03	2.48	1.67	1414.3	3.9
LATITUDE +15:		1.97	2.67	3.42	4.00	4.34	4.59	4.66	4.86	4.63	4.07	2.58	1.73	1324.9	3.6
DODGE CITY	KS	LATITUDE:		37 DEGREES	46 MINUTES										
LATITUDE -15:		3.67	4.46	5.31	6.20	6.47	7.14	7.11	6.79	6.14	5.31	4.05	3.47	2012.6	5.5
LATITUDE :		4.15	4.78	5.40	5.97	5.98	6.48	6.55	6.54	6.27	5.76	4.60	4.01	2023.5	5.5
LATITUDE +15:		4.38	4.84	5.21	5.45	5.24	5.57	5.72	5.97	6.07	5.88	4.88	4.31	1933.0	5.3

The Solar Electric Independent Home Book

SITE ARRAY TILT		JAN	FEB	MAR	APR	MAY	JUN	JUL	AUG	SEP	OCT	NOV	DEC	ANNUAL TOTAL (KWH/SQ. M)	AVERAGE DAY (KWH/SQ. M)
GOODLAND	KS	LATITUDE:	39 DEGREES	22 MINUTES											
LATITUDE -15:		3.67	4.28	5.19	6.05	6.39	7.13	7.19	6.81	6.07	5.34	4.07	3.47	1999.4	5.5
LATITUDE :		4.15	4.58	5.27	5.82	5.90	6.46	6.62	6.56	6.20	5.80	4.63	4.01	2008.8	5.5
LATITUDE +15:		4.38	4.63	5.08	5.31	5.17	5.54	5.77	5.98	5.98	5.92	4.92	4.32	1917.5	5.3
TOPEKA	KS	LATITUDE:	39 DEGREES	4 MINUTES											
LATITUDE -15:		3.03	3.72	4.50	5.39	5.93	6.44	6.59	6.32	5.54	4.72	3.55	2.76	1781.4	4.9
LATITUDE :		3.38	3.95	4.55	5.18	5.48	5.85	6.08	6.08	5.64	5.08	4.00	3.15	1778.0	4.9
LATITUDE +15:		3.55	3.98	4.37	4.73	4.81	5.04	5.31	5.54	5.43	5.16	4.22	3.36	1688.6	4.6
WICHITA	KS	LATITUDE:	37 DEGREES	39 MINUTES											
LATITUDE -15:		3.43	4.15	5.01	5.84	6.30	6.86	6.93	6.71	5.84	5.06	3.92	3.21	1926.1	5.3
LATITUDE :		3.85	4.43	5.09	5.62	5.82	6.23	6.39	6.47	5.96	5.47	4.44	3.69	1931.4	5.3
LATITUDE +15:		4.05	4.48	4.90	5.14	5.11	5.36	5.58	5.90	5.75	5.57	4.70	3.95	1841.3	5.0
LEXINGTON	KY	LATITUDE:	38 DEGREES	2 MINUTES											
LATITUDE -15:		2.25	2.95	3.85	4.81	5.39	5.74	5.72	5.52	4.85	4.15	2.83	2.12	1528.7	4.2
LATITUDE :		2.46	3.10	3.88	4.62	4.99	5.23	5.28	5.30	4.91	4.43	3.14	2.37	1513.9	4.1
LATITUDE +15:		2.54	3.09	3.71	4.22	4.40	4.54	4.64	4.84	4.72	4.47	3.29	2.50	1429.3	3.9
LOUISVILLE	KY	LATITUDE:	38 DEGREES	11 MINUTES											
LATITUDE -15:		2.26	3.00	3.87	4.78	5.31	5.76	5.67	5.51	4.86	4.14	2.82	2.14	1526.7	4.2
LATITUDE :		2.47	3.15	3.89	4.59	4.92	5.26	5.24	5.29	4.92	4.42	3.13	2.40	1512.8	4.1
LATITUDE +15:		2.55	3.14	3.73	4.19	4.34	4.56	4.61	4.83	4.73	4.47	3.27	2.53	1429.0	3.9
BATON ROUGE	LA	LATITUDE:	30 DEGREES	32 MINUTES											
LATITUDE -15:		2.92	3.74	4.64	5.38	5.77	5.86	5.38	5.37	4.93	4.71	3.48	2.83	1675.1	4.6
LATITUDE :		3.22	3.96	4.70	5.19	5.36	5.37	5.01	5.18	5.00	5.06	3.88	3.19	1677.2	4.6
LATITUDE +15:		3.35	3.98	4.53	4.76	4.74	4.67	4.43	4.75	4.81	5.13	4.07	3.39	1600.1	4.4
LAKE CHARLES	LA	LATITUDE:	30 DEGREES	7 MINUTES											
LATITUDE -15:		2.67	3.56	4.39	5.01	5.71	5.99	5.51	5.29	4.99	5.01	3.45	2.67	1651.9	4.5
LATITUDE :		2.91	3.75	4.44	4.83	5.31	5.48	5.12	5.11	5.06	5.39	3.84	3.00	1650.6	4.5
LATITUDE +15:		3.01	3.76	4.27	4.43	4.69	4.75	4.53	4.68	4.87	5.47	4.02	3.16	1571.8	4.3
NEW ORLEANS	LA	LATITUDE:	29 DEGREES	59 MINUTES											
LATITUDE -15:		3.09	3.93	4.74	5.70	6.08	6.11	5.60	5.49	5.09	4.81	3.66	2.98	1743.3	4.8
LATITUDE :		3.41	4.17	4.81	5.50	5.65	5.59	5.21	5.30	5.17	5.17	4.09	3.37	1747.9	4.8
LATITUDE +15:		3.56	4.20	4.63	5.05	4.98	4.85	4.61	4.86	4.99	5.25	4.30	3.58	1669.2	4.6
SHREVEPORT	LA	LATITUDE:	32 DEGREES	28 MINUTES											
LATITUDE -15:		2.93	3.77	4.56	5.18	5.82	6.27	6.20	6.05	5.33	4.85	3.67	2.94	1753.2	4.8
LATITUDE :		3.23	3.99	4.62	4.99	5.40	5.72	5.74	5.84	5.42	5.22	4.11	3.33	1753.8	4.8
LATITUDE +15:		3.37	4.01	4.44	4.57	4.76	4.95	5.05	5.34	5.22	5.30	4.32	3.54	1670.6	4.6
BOSTON	MA	LATITUDE:	42 DEGREES	22 MINUTES											
LATITUDE -15:		2.17	2.86	3.69	4.37	5.02	5.50	5.43	4.93	4.67	3.74	2.34	1.97	1421.2	3.9
LATITUDE :		2.38	3.00	3.71	4.18	4.64	5.01	5.01	4.73	4.72	3.99	2.59	2.21	1405.3	3.9
LATITUDE +15:		2.47	2.99	3.54	3.81	4.08	4.34	4.40	4.31	4.53	4.02	2.70	2.33	1325.0	3.6
BALTIMORE	MD	LATITUDE:	39 DEGREES	11 MINUTES											
LATITUDE -15:		2.55	3.28	4.13	4.87	5.29	5.69	5.63	5.25	4.79	4.01	2.94	2.29	1544.2	4.2
LATITUDE :		2.81	3.47	4.17	4.67	4.90	5.19	5.20	5.04	4.84	4.28	3.28	2.58	1534.8	4.2
LATITUDE +15:		2.93	3.47	3.99	4.26	4.31	4.50	4.57	4.60	4.65	4.32	3.44	2.73	1453.9	4.0
PATUXENT RIVER	MD	LATITUDE:	38 DEGREES	17 MINUTES											
LATITUDE -15:		2.59	3.33	4.17	5.02	5.45	5.73	5.61	5.33	4.85	4.05	3.10	2.42	1572.7	4.3
LATITUDE :		2.86	3.52	4.21	4.82	5.04	5.23	5.19	5.12	4.91	4.32	3.47	2.74	1565.0	4.3
LATITUDE +15:		2.98	3.53	4.04	4.40	4.44	4.53	4.56	4.67	4.72	4.36	3.64	2.91	1484.2	4.1
BANGOR	ME	LATITUDE:	44 DEGREES	48 MINUTES											
LATITUDE -15:		2.25	3.09	4.11	4.83	5.39	5.63	5.80	5.45	4.80	3.69	2.36	2.02	1504.8	4.1
LATITUDE :		2.49	3.27	4.15	4.62	4.97	5.12	5.34	5.22	4.86	3.95	2.63	2.29	1489.3	4.1
LATITUDE +15:		2.60	3.27	3.98	4.21	4.36	4.43	4.68	4.76	4.67	3.98	2.75	2.43	1404.1	3.8
CARIBOU	ME	LATITUDE:	46 DEGREES	52 MINUTES											
LATITUDE -15:		2.21	3.24	4.40	4.78	4.93	5.33	5.51	5.10	4.26	3.07	1.85	1.74	1413.0	3.9
LATITUDE :		2.46	3.44	4.46	4.58	4.54	4.84	5.07	4.88	4.30	3.27	2.03	1.97	1394.9	3.8
LATITUDE +15:		2.57	3.46	4.28	4.16	3.99	4.19	4.44	4.44	4.12	3.28	2.11	2.08	1312.1	3.6
PORTLAND	ME	LATITUDE:	43 DEGREES	39 MINUTES											
LATITUDE -15:		2.12	2.79	3.54	4.31	4.86	5.18	5.15	4.87	4.32	3.51	2.18	1.81	1359.8	3.7
LATITUDE :		2.33	2.93	3.56	4.13	4.49	4.72	4.75	4.67	4.36	3.74	2.41	2.03	1343.3	3.7
LATITUDE +15:		2.43	2.93	3.40	3.76	3.95	4.10	4.17	4.25	4.18	3.77	2.51	2.14	1265.6	3.5
ALPENA	MI	LATITUDE:	45 DEGREES	4 MINUTES											
LATITUDE -15:		1.67	2.55	3.85	4.71	5.36	5.69	5.87	5.35	4.39	3.23	1.81	1.49	1400.8	3.8
LATITUDE :		1.82	2.66	3.88	4.50	4.94	5.16	5.40	5.12	4.43	3.42	1.98	1.70	1371.6	3.8
LATITUDE +15:		1.87	2.65	3.71	4.10	4.33	4.46	4.73	4.66	4.24	3.44	2.04	1.81	1280.1	3.5

SITE ARRAY TILT		JAN	FEB	MAR	APR	MAY	JUN	JUL	AUG	SEP	OCT	NOV	DEC	ANNUAL TOTAL (KWH/SQ. M)	AVERAGE DAY (KWH/SQ. M)
DETROIT	MI	LATITUDE:		42 DEGREES	25 MINUTES										
LATITUDE -15:		1.83	2.71	3.61	4.63	5.32	5.66	5.70	5.24	4.65	3.69	2.19	1.58	1425.7	3.9
LATITUDE :		1.99	2.84	3.63	4.43	4.92	5.15	5.26	5.02	4.70	3.93	2.42	1.75	1401.6	3.8
LATITUDE +15:		2.05	2.83	3.47	4.04	4.32	4.46	4.61	4.58	4.51	3.96	2.51	1.83	1313.9	3.6
FLINT	MI	LATITUDE:		42 DEGREES	58 MINUTES										
LATITUDE -15:		1.65	2.53	3.46	4.42	5.15	5.49	5.59	5.19	4.43	3.49	1.94	1.60	1369.4	3.8
LATITUDE :		1.79	2.64	3.47	4.24	4.76	5.00	5.16	4.98	4.48	3.72	2.13	1.82	1345.7	3.7
LATITUDE +15:		1.83	2.62	3.31	3.86	4.19	4.34	4.53	4.54	4.30	3.75	2.20	1.94	1261.1	3.5
GRAND RAPIDS	MI	LATITUDE:		42 DEGREES	53 MINUTES										
LATITUDE -15:		1.78	2.57	3.69	4.68	5.46	5.93	5.95	5.62	4.71	3.63	2.03	1.60	1452.3	4.0
LATITUDE :		1.98	2.69	3.71	4.48	5.04	5.40	5.49	5.39	4.77	3.87	2.23	1.83	1428.6	3.9
LATITUDE +15:		2.08	2.68	3.55	4.09	4.43	4.67	4.81	4.92	4.58	3.90	2.32	1.95	1339.2	3.7
HOUGHTON	MI	LATITUDE:		47 DEGREES	10 MINUTES										
LATITUDE -15:		1.23	1.96	3.54	4.61	5.18	5.56	5.76	5.19	3.88	3.01	1.58	1.05	1297.0	3.6
LATITUDE :		1.35	2.02	3.56	4.40	4.77	5.04	5.29	4.96	3.90	3.19	1.76	1.17	1262.7	3.5
LATITUDE +15:		1.40	1.98	3.40	4.00	4.18	4.36	4.62	4.51	3.73	3.20	1.84	1.23	1172.0	3.2
SAULT STE. MARIE	MI	LATITUDE:		46 DEGREES	28 MINUTES										
LATITUDE -15:		1.52	2.56	3.93	4.66	5.26	5.49	5.74	5.16	4.00	2.96	1.58	1.48	1351.2	3.7
LATITUDE :		1.64	2.69	3.96	4.45	4.85	4.98	5.28	4.94	4.03	3.14	1.72	1.69	1321.4	3.6
LATITUDE +15:		1.69	2.68	3.79	4.05	4.25	4.31	4.62	4.49	3.86	3.15	1.76	1.80	1231.9	3.4
TRAVERSE CITY	MI	LATITUDE:		44 DEGREES	44 MINUTES										
LATITUDE -15:		1.53	2.27	3.72	4.69	5.39	5.80	5.96	5.44	4.40	3.25	1.74	1.37	1388.7	3.8
LATITUDE :		1.70	2.36	3.74	4.49	4.97	5.27	5.49	5.21	4.45	3.46	1.90	1.55	1358.7	3.7
LATITUDE +15:		1.77	2.33	3.58	4.09	4.36	4.55	4.81	4.75	4.26	3.47	1.96	1.64	1267.1	3.5
DULUTH	MN	LATITUDE:		46 DEGREES	50 MINUTES										
LATITUDE -15:		1.99	2.97	3.97	4.63	5.13	5.35	5.81	5.27	4.23	3.27	1.95	1.59	1405.7	3.9
LATITUDE :		2.20	3.14	4.00	4.43	4.73	4.86	5.34	5.05	4.27	3.49	2.15	1.79	1383.8	3.8
LATITUDE +15:		2.29	3.14	3.83	4.03	4.15	4.21	4.68	4.59	4.09	3.51	2.24	1.89	1298.1	3.6
INTERNATIONAL FALL	MN	LATITUDE:		48 DEGREES	34 MINUTES										
LATITUDE -15:		1.93	3.05	4.13	4.94	5.38	5.62	6.04	5.60	4.45	3.32	1.86	1.63	1460.5	4.0
LATITUDE :		2.13	3.24	4.17	4.73	4.95	5.10	5.55	5.36	4.51	3.55	2.06	1.84	1436.4	3.9
LATITUDE +15:		2.22	3.25	3.99	4.29	4.33	4.40	4.85	4.88	4.32	3.57	2.14	1.95	1345.2	3.7
MINNEAPOLIS-ST. PAUL	MN	LATITUDE:		44 DEGREES	53 MINUTES										
LATITUDE -15:		2.31	3.29	4.16	4.84	5.41	5.84	6.16	5.73	4.81	3.81	2.43	1.85	1541.8	4.2
LATITUDE :		2.57	3.49	4.20	4.63	4.99	5.31	5.67	5.50	4.87	4.08	2.71	2.09	1525.4	4.2
LATITUDE +15:		2.69	3.50	4.03	4.22	4.38	4.59	4.96	5.01	4.68	4.12	2.83	2.21	1437.2	3.9
ROCHESTER	MN	LATITUDE:		43 DEGREES	55 MINUTES										
LATITUDE -15:		2.31	3.16	4.02	4.70	5.28	5.76	5.95	5.60	4.72	3.77	2.42	1.88	1510.0	4.1
LATITUDE :		2.56	3.35	4.06	4.50	4.87	5.24	5.49	5.38	4.79	4.03	2.70	2.12	1494.0	4.1
LATITUDE +15:		2.67	3.36	3.89	4.10	4.28	4.54	4.81	4.90	4.60	4.07	2.82	2.24	1408.3	3.9
COLUMBIA	MO	LATITUDE:		38 DEGREES	49 MINUTES										
LATITUDE -15:		2.64	3.41	4.19	4.99	5.82	6.33	6.55	6.21	5.24	4.47	3.13	2.38	1686.8	4.6
LATITUDE :		2.93	3.61	4.23	4.79	5.39	5.76	6.05	5.98	5.33	4.80	3.50	2.69	1676.9	4.6
LATITUDE +15:		3.05	3.62	4.06	4.38	4.74	4.98	5.30	5.46	5.13	4.87	3.68	2.86	1587.2	4.3
KANSAS CITY	MO	LATITUDE:		39 DEGREES	18 MINUTES										
LATITUDE -15:		2.88	3.53	4.30	5.17	5.80	6.30	6.51	6.16	5.29	4.48	3.37	2.66	1718.9	4.7
LATITUDE :		3.21	3.74	4.34	4.96	5.36	5.72	6.00	5.93	5.37	4.81	3.78	3.02	1712.9	4.7
LATITUDE +15:		3.36	3.76	4.17	4.53	4.71	4.94	5.25	5.41	5.17	4.88	3.99	3.22	1624.9	4.5
SPRINGFIELD	MO	LATITUDE:		37 DEGREES	14 MINUTES										
LATITUDE -15:		2.89	3.55	4.34	5.23	5.82	6.28	6.37	6.14	5.28	4.53	3.36	2.70	1720.5	4.7
LATITUDE :		3.21	3.76	4.38	5.02	5.38	5.72	5.88	5.91	5.36	4.87	3.76	3.06	1714.8	4.7
LATITUDE +15:		3.36	3.77	4.21	4.59	4.73	4.94	5.15	5.40	5.16	4.94	3.95	3.26	1627.2	4.5
ST. LOUIS	MO	LATITUDE:		38 DEGREES	45 MINUTES										
LATITUDE -15:		2.72	3.46	4.28	5.12	5.79	6.34	6.34	5.99	5.28	4.45	3.20	2.42	1687.1	4.6
LATITUDE :		3.02	3.66	4.33	4.92	5.36	5.77	5.85	5.76	5.37	4.79	3.59	2.74	1679.1	4.6
LATITUDE +15:		3.16	3.68	4.15	4.50	4.71	4.99	5.13	5.26	5.17	4.85	3.77	2.91	1591.1	4.4
JACKSON	MS	LATITUDE:		32 DEGREES	19 MINUTES										
LATITUDE -15:		2.88	3.71	4.66	5.49	5.98	6.14	5.88	5.73	5.16	4.72	3.54	2.83	1726.7	4.7
LATITUDE :		3.17	3.93	4.72	5.29	5.55	5.61	5.45	5.52	5.23	5.06	3.95	3.20	1725.1	4.7
LATITUDE +15:		3.30	3.94	4.54	4.84	4.89	4.86	4.80	5.05	5.04	5.14	4.14	3.39	1641.5	4.5
MERIDIAN	MS	LATITUDE:		32 DEGREES	20 MINUTES										
LATITUDE -15:		2.84	3.66	4.51	5.33	5.73	5.96	5.62	5.60	4.96	4.66	3.51	2.79	1679.7	4.6
LATITUDE :		3.13	3.88	4.57	5.14	5.32	5.45	5.21	5.39	5.03	5.00	3.92	3.15	1679.2	4.6
LATITUDE +15:		3.25	3.89	4.39	4.71	4.69	4.73	4.60	4.94	4.84	5.07	4.11	3.34	1599.3	4.4

The Solar Electric Independent Home Book

SITE ARRAY TILT		JAN	FEB	MAR	APR	MAY	JUN	JUL	AUG	SEP	OCT	NOV	DEC	ANNUAL TOTAL (KWH/SQ. M)	AVERAGE DAY (KWH/SQ. M)
BILLINGS	MT	LATITUDE:		45 DEGREES	48 MINUTES										
LATITUDE -15:		2.55	3.36	4.59	5.15	5.99	6.61	7.51	7.02	5.87	4.65	3.08	2.45	1792.7	4.9
LATITUDE :		2.85	3.57	4.65	4.93	5.52	5.99	6.90	6.76	5.99	5.02	3.48	2.81	1781.6	4.9
LATITUDE +15:		2.99	3.59	4.47	4.49	4.83	5.15	6.01	6.16	5.79	5.11	3.67	3.00	1683.0	4.6
CUT BANK	MT	LATITUDE:		48 DEGREES	36 MINUTES										
LATITUDE -15:		2.28	3.20	4.50	5.10	5.93	6.22	7.25	6.68	5.57	4.36	2.89	2.17	1711.5	4.7
LATITUDE :		2.55	3.40	4.56	4.87	5.45	5.63	6.65	6.42	5.68	4.71	3.26	2.49	1696.8	4.6
LATITUDE +15:		2.67	3.42	4.38	4.43	4.77	4.84	5.79	5.85	5.47	4.78	3.44	2.66	1599.3	4.4
DILLON	MT	LATITUDE:		45 DEGREES	15 MINUTES										
LATITUDE -15:		2.76	3.75	4.96	5.56	6.23	6.50	7.53	6.99	6.05	4.79	3.30	2.59	1858.4	5.1
LATITUDE :		3.09	4.00	5.04	5.32	5.73	5.89	6.91	6.72	6.18	5.19	3.73	2.97	1850.8	5.1
LATITUDE +15:		3.25	4.03	4.85	4.85	5.01	5.06	6.01	6.12	5.97	5.28	3.94	3.18	1751.6	4.8
GLASGOW	MT	LATITUDE:		48 DEGREES	13 MINUTES										
LATITUDE -15:		2.13	3.08	4.37	5.09	5.73	6.22	6.93	6.53	5.48	4.34	2.82	2.12	1671.4	4.6
LATITUDE :		2.37	3.26	4.42	4.87	5.27	5.62	6.36	6.27	5.58	4.68	3.17	2.43	1654.7	4.5
LATITUDE +15:		2.48	3.27	4.24	4.42	4.61	4.84	5.54	5.71	5.38	4.75	3.34	2.59	1557.9	4.3
GREAT FALLS	MT	LATITUDE:		47 DEGREES	29 MINUTES										
LATITUDE -15:		2.28	3.28	4.62	5.07	5.80	6.39	7.36	6.76	5.60	4.52	2.85	2.02	1723.5	4.7
LATITUDE :		2.55	3.48	4.68	4.85	5.33	5.79	6.76	6.50	5.71	4.88	3.21	2.31	1707.5	4.7
LATITUDE +15:		2.66	3.50	4.50	4.41	4.67	4.97	5.88	5.92	5.50	4.96	3.38	2.46	1608.4	4.4
HELENA	MT	LATITUDE:		46 DEGREES	36 MINUTES										
LATITUDE -15:		2.18	3.13	4.44	5.04	5.82	6.20	7.36	6.70	5.67	4.41	2.90	2.13	1706.7	4.7
LATITUDE :		2.43	3.32	4.50	4.82	5.36	5.62	6.76	6.44	5.79	4.76	3.26	2.43	1690.7	4.6
LATITUDE +15:		2.53	3.33	4.32	4.39	4.69	4.84	5.88	5.87	5.58	4.83	3.44	2.59	1592.8	4.4
LEWISTOWN	MT	LATITUDE:		47 DEGREES	3 MINUTES										
LATITUDE -15:		2.24	3.09	4.40	4.90	5.66	6.25	7.22	6.62	5.52	4.35	2.83	2.19	1684.1	4.6
LATITUDE :		2.49	3.27	4.45	4.68	5.20	5.66	6.62	6.35	5.62	4.69	3.18	2.50	1667.2	4.6
LATITUDE +15:		2.60	3.28	4.27	4.26	4.55	4.86	5.76	5.78	5.41	4.76	3.35	2.66	1569.7	4.3
MILES CITY	MT	LATITUDE:		46 DEGREES	26 MINUTES										
LATITUDE -15:		2.43	3.32	4.61	5.23	5.94	6.52	7.22	6.88	5.80	4.59	3.10	2.38	1767.4	4.8
LATITUDE :		2.71	3.53	4.67	5.01	5.47	5.90	6.63	6.61	5.92	4.96	3.51	2.72	1755.3	4.8
LATITUDE +15:		2.84	3.54	4.48	4.56	4.78	5.07	5.77	6.02	5.71	5.04	3.70	2.91	1657.2	4.5
MISSOULA	MT	LATITUDE:		46 DEGREES	55 MINUTES										
LATITUDE -15:		1.46	2.43	3.74	4.67	5.58	5.87	7.35	6.54	5.43	3.77	2.16	1.41	1537.4	4.2
LATITUDE :		1.58	2.55	3.77	4.47	5.14	5.33	6.76	6.28	5.54	4.04	2.40	1.57	1506.6	4.1
LATITUDE +15:		1.62	2.53	3.60	4.06	4.51	4.60	5.88	5.72	5.34	4.08	2.51	1.64	1405.2	3.8
ASHEVILLE	NC	LATITUDE:		35 DEGREES	26 MINUTES										
LATITUDE -15:		2.93	3.65	4.54	5.41	5.57	5.63	5.47	5.27	4.73	4.40	3.56	2.82	1643.1	4.5
LATITUDE :		3.25	3.86	4.59	5.20	5.16	5.15	5.07	5.07	4.79	4.72	3.99	3.20	1645.1	4.5
LATITUDE +15:		3.40	3.88	4.42	4.76	4.55	4.48	4.47	4.64	4.60	4.78	4.21	3.41	1569.4	4.3
CAPE HATTERAS	NC	LATITUDE:		35 DEGREES	16 MINUTES										
LATITUDE -15:		2.75	3.55	4.61	5.76	6.05	6.17	5.92	5.53	5.14	4.35	3.66	2.82	1714.8	4.7
LATITUDE :		3.04	3.76	4.66	5.55	5.60	5.62	5.48	5.32	5.22	4.66	4.11	3.20	1711.1	4.7
LATITUDE +15:		3.17	3.77	4.48	5.07	4.93	4.87	4.82	4.87	5.02	4.72	4.34	3.40	1625.9	4.5
CHARLOTTE	NC	LATITUDE:		35 DEGREES	13 MINUTES										
LATITUDE -15:		2.91	3.64	4.57	5.50	5.72	5.83	5.64	5.50	4.93	4.51	3.62	2.88	1681.8	4.6
LATITUDE :		3.22	3.85	4.62	5.29	5.30	5.32	5.23	5.29	5.00	4.83	4.06	3.28	1682.8	4.6
LATITUDE +15:		3.36	3.87	4.44	4.84	4.67	4.62	4.60	4.84	4.81	4.90	4.28	3.49	1603.7	4.4
CHERRY POINT	NC	LATITUDE:		34 DEGREES	54 MINUTES										
LATITUDE -15:		3.06	3.83	4.82	5.82	5.94	5.89	5.64	5.29	4.96	4.46	3.78	3.07	1721.2	4.7
LATITUDE :		3.40	4.07	4.88	5.61	5.50	5.38	5.23	5.09	5.03	4.79	4.26	3.51	1726.8	4.7
LATITUDE +15:		3.56	4.10	4.70	5.13	4.85	4.67	4.61	4.66	4.84	4.85	4.51	3.75	1649.7	4.5
GREENSBORO	NC	LATITUDE:		36 DEGREES	5 MINUTES										
LATITUDE -15:		2.95	3.67	4.59	5.47	5.76	5.92	5.75	5.52	4.98	4.43	3.58	2.89	1689.5	4.6
LATITUDE :		3.28	3.89	4.64	5.26	5.33	5.40	5.32	5.30	5.05	4.75	4.02	3.28	1689.4	4.6
LATITUDE +15:		3.42	3.91	4.46	4.80	4.69	4.68	4.67	4.84	4.85	4.81	4.23	3.50	1608.9	4.4
RALEIGH-DURHAM	NC	LATITUDE:		35 DEGREES	52 MINUTES										
LATITUDE -15:		2.83	3.55	4.44	5.34	5.58	5.66	5.47	5.22	4.81	4.25	3.40	2.74	1622.3	4.4
LATITUDE :		3.13	3.75	4.49	5.14	5.18	5.17	5.07	5.03	4.87	4.55	3.81	3.11	1622.5	4.4
LATITUDE +15:		3.27	3.77	4.32	4.70	4.57	4.50	4.48	4.60	4.69	4.60	4.01	3.31	1546.1	4.2
BISMARCK	ND	LATITUDE:		46 DEGREES	46 MINUTES										
LATITUDE -15:		2.53	3.51	4.55	4.94	5.79	6.26	6.88	6.52	5.40	4.31	2.82	2.22	1697.8	4.7
LATITUDE :		2.83	3.75	4.61	4.73	5.33	5.67	6.32	6.26	5.51	4.65	3.18	2.54	1686.3	4.6
LATITUDE +15:		2.97	3.77	4.43	4.30	4.67	4.88	5.51	5.70	5.30	4.72	3.35	2.71	1592.8	4.4

SITE ARRAY TILT		JAN	FEB	MAR	APR	MAY	JUN	JUL	AUG	SEP	OCT	NOV	DEC	ANNUAL TOTAL (KWH/SQ. M)	AVERAGE DAY (KWH/SQ. M)
FARGO	ND	LATITUDE:	46 DEGREES	54 MINUTES											
LATITUDE -15:		2.18	3.15	4.25	5.00	5.75	6.05	6.67	6.32	5.19	4.13	2.49	1.95	1618.6	4.4
LATITUDE :		2.42	3.34	4.30	4.79	5.29	5.49	6.13	6.07	5.28	4.44	2.79	2.22	1601.1	4.4
LATITUDE +15:		2.53	3.35	4.12	4.36	4.64	4.73	5.35	5.53	5.08	4.51	2.93	2.36	1507.0	4.1
MINOT	ND	LATITUDE:	48 DEGREES	16 MINUTES											
LATITUDE -15:		2.11	2.99	4.10	4.99	5.80	6.00	6.62	6.29	5.18	4.18	2.52	1.92	1606.1	4.4
LATITUDE :		2.34	3.17	4.14	4.77	5.34	5.43	6.07	6.03	5.26	4.50	2.83	2.19	1586.7	4.3
LATITUDE +15:		2.45	3.18	3.97	4.33	4.66	4.68	5.29	5.49	5.06	4.56	2.97	2.33	1491.4	4.1
GRAND ISLAND	NE	LATITUDE:	40 DEGREES	58 MINUTES											
LATITUDE -15:		3.10	3.75	4.62	5.62	6.13	6.80	6.89	6.50	5.63	4.86	3.58	2.86	1837.6	5.0
LATITUDE :		3.48	3.99	4.68	5.40	5.66	6.17	6.35	6.26	5.74	5.26	4.05	3.28	1836.5	5.0
LATITUDE +15:		3.66	4.02	4.50	4.93	4.97	5.31	5.55	5.71	5.54	5.35	4.29	3.52	1745.5	4.8
NORTH OMAHA	NE	LATITUDE:	41 DEGREES	22 MINUTES											
LATITUDE -15:		3.00	3.66	4.48	5.16	5.81	6.43	6.55	6.22	5.08	4.46	3.06	2.56	1719.8	4.7
LATITUDE :		3.36	3.89	4.53	4.95	5.36	5.84	6.03	5.98	5.16	4.80	3.43	2.91	1712.3	4.7
LATITUDE +15:		3.53	3.91	4.35	4.51	4.71	5.03	5.27	5.45	4.96	4.87	3.61	3.11	1622.5	4.4
NORTH PLATTE	NE	LATITUDE:	41 DEGREES	8 MINUTES											
LATITUDE -15:		3.31	3.95	4.91	5.73	6.17	6.87	7.09	6.67	5.88	5.09	3.74	3.13	1905.3	5.2
LATITUDE :		3.72	4.22	4.98	5.50	5.69	6.22	6.52	6.42	6.00	5.51	4.24	3.61	1907.1	5.2
LATITUDE +15:		3.92	4.25	4.79	5.01	4.98	5.35	5.68	5.85	5.79	5.62	4.49	3.87	1814.5	5.0
SCOTTSBLUFF	NE	LATITUDE:	41 DEGREES	52 MINUTES											
LATITUDE -15:		3.29	3.98	4.85	5.55	6.02	6.78	7.12	6.75	6.08	5.01	3.61	3.02	1889.5	5.2
LATITUDE :		3.71	4.25	4.92	5.33	5.55	6.15	6.55	6.50	6.22	5.42	4.09	3.48	1892.9	5.2
LATITUDE +15:		3.91	4.29	4.74	4.87	4.87	5.29	5.72	5.93	6.01	5.53	4.33	3.73	1802.4	4.9
CONCORD	NH	LATITUDE:	43 DEGREES	12 MINUTES											
LATITUDE -15:		2.14	2.78	3.54	4.35	4.90	5.16	5.19	4.83	4.22	3.45	2.17	1.77	1355.8	3.7
LATITUDE :		2.36	2.92	3.56	4.16	4.53	4.70	4.79	4.63	4.25	3.67	2.39	1.98	1337.6	3.7
LATITUDE +15:		2.45	2.91	3.40	3.79	3.98	4.08	4.20	4.21	4.07	3.69	2.49	2.08	1258.8	3.4
LAKEHURST	NJ	LATITUDE:	40 DEGREES	2 MINUTES											
LATITUDE -15:		2.47	3.13	3.96	4.78	5.17	5.38	5.26	5.04	4.55	3.88	2.81	2.22	1480.9	4.1
LATITUDE :		2.73	3.30	3.99	4.58	4.78	4.90	4.86	4.83	4.60	4.13	3.13	2.50	1470.9	4.0
LATITUDE +15:		2.84	3.30	3.82	4.18	4.21	4.26	4.28	4.41	4.41	4.17	3.28	2.65	1392.8	3.8
NEWARK	NJ	LATITUDE:	40 DEGREES	42 MINUTES											
LATITUDE -15:		2.47	3.15	3.98	4.76	5.23	5.44	5.46	5.16	4.63	3.90	2.72	2.14	1493.2	4.1
LATITUDE :		2.73	3.32	4.01	4.57	4.84	4.96	5.04	4.95	4.69	4.16	3.03	2.41	1483.1	4.1
LATITUDE +15:		2.85	3.33	3.85	4.17	4.26	4.31	4.43	4.52	4.50	4.20	3.18	2.55	1404.2	3.8
ALBUQUERQUE	NM	LATITUDE:	35 DEGREES	3 MINUTES											
LATITUDE -15:		4.35	5.22	6.28	7.29	7.83	8.09	7.67	7.50	7.07	6.17	4.98	4.23	2334.2	6.4
LATITUDE :		4.94	5.62	6.43	7.03	7.21	7.30	7.06	7.24	7.26	6.73	5.70	4.92	2356.8	6.5
LATITUDE +15:		5.24	5.72	6.22	6.42	6.27	6.22	6.14	6.60	7.03	6.90	6.09	5.33	2257.0	6.2
CLAYTON	NM	LATITUDE:	36 DEGREES	27 MINUTES											
LATITUDE -15:		4.24	4.88	5.92	6.69	6.86	7.31	7.05	6.89	6.51	5.80	4.62	4.04	2155.5	5.9
LATITUDE :		4.81	5.25	6.05	6.44	6.34	6.63	6.50	6.64	6.66	6.31	5.28	4.71	2179.6	6.0
LATITUDE +15:		5.11	5.33	5.84	5.88	5.55	5.69	5.68	6.06	6.45	6.46	5.63	5.10	2092.3	5.7
FARMINGTON	NM	LATITUDE:	36 DEGREES	45 MINUTES											
LATITUDE -15:		4.19	5.09	6.10	7.02	7.59	8.06	7.66	7.44	7.05	6.04	4.76	3.94	2280.6	6.2
LATITUDE :		4.75	5.49	6.24	6.77	6.99	7.29	7.05	7.18	7.25	6.59	5.45	4.59	2301.7	6.3
LATITUDE +15:		5.05	5.59	6.03	6.18	6.10	6.22	6.14	6.55	7.03	6.76	5.83	4.96	2203.8	6.0
ROSWELL	NM	LATITUDE:	33 DEGREES	24 MINUTES											
LATITUDE -15:		4.30	5.20	6.33	7.21	7.58	7.90	7.52	7.29	6.73	5.88	4.73	4.12	2276.3	6.2
LATITUDE :		4.87	5.61	6.48	6.96	6.99	7.14	6.93	7.03	6.89	6.40	5.40	4.78	2297.1	6.3
LATITUDE +15:		5.16	5.71	6.27	6.35	6.10	6.10	6.04	6.42	6.67	6.55	5.76	5.17	2199.6	6.0
TRUTH OR CONSEQUEN	NM	LATITUDE:	33 DEGREES	14 MINUTES											
LATITUDE -15:		4.64	5.53	6.62	7.60	7.89	8.01	7.28	7.20	6.82	6.10	5.13	4.37	2348.5	6.4
LATITUDE :		5.28	5.98	6.79	7.33	7.27	7.24	6.72	6.95	6.98	6.64	5.88	5.09	2377.3	6.5
LATITUDE +15:		5.61	6.09	6.58	6.69	6.32	6.18	5.86	6.34	6.76	6.81	6.29	5.52	2282.5	6.3
TUCUMCARI	NM	LATITUDE:	35 DEGREES	11 MINUTES											
LATITUDE -15:		4.33	5.02	6.08	6.86	7.14	7.51	7.24	7.08	6.53	5.70	4.67	4.14	2201.1	6.0
LATITUDE :		4.91	5.41	6.21	6.61	6.59	6.80	6.67	6.83	6.68	6.20	5.34	4.82	2223.5	6.1
LATITUDE +15:		5.21	5.49	6.00	6.04	5.76	5.82	5.82	6.23	6.46	6.34	5.69	5.22	2132.1	5.8
ZUNI	NM	LATITUDE:	35 DEGREES	6 MINUTES											
LATITUDE -15:		4.20	5.02	5.98	7.08	7.63	7.86	6.98	6.79	6.78	5.93	4.75	4.04	2223.0	6.1
LATITUDE :		4.76	5.40	6.11	6.83	7.03	7.11	6.44	6.54	6.94	6.46	5.43	4.69	2243.3	6.1
LATITUDE +15:		5.05	5.49	5.90	6.23	6.13	6.07	5.63	5.97	5.72	6.61	5.79	5.07	2148.6	5.9

The Solar Electric Independent Home Book

SITE ARRAY TILT		JAN	FEB	MAR	APR	MAY	JUN	JUL	AUG	SEP	OCT	NOV	DEC	ANNUAL TOTAL (KWH/SQ. M)	AVERAGE DAY (KWH/SQ. M)
ELKO	NV	LATITUDE: 40 DEGREES 50 MINUTES													
LATITUDE -15:		3.25	4.31	5.44	6.34	7.17	7.67	8.18	7.84	7.28	5.80	4.01	3.16	2145.9	5.9
LATITUDE :		3.65	4.62	5.55	6.10	6.61	6.94	7.52	7.57	7.50	6.33	4.57	3.65	2149.4	5.9
LATITUDE +15:		3.85	4.67	5.35	5.57	5.77	5.94	6.53	6.90	7.28	6.49	4.86	3.92	2043.2	5.6
ELY	NV	LATITUDE: 39 DEGREES 17 MINUTES													
LATITUDE -15:		3.84	4.69	5.93	6.68	7.17	7.61	7.60	7.46	7.31	6.04	4.48	3.63	2204.7	6.0
LATITUDE :		4.34	5.04	6.06	6.43	6.60	6.88	6.99	7.19	7.52	6.59	5.12	4.21	2220.7	6.1
LATITUDE +15:		4.60	5.11	5.85	5.87	5.76	5.89	6.08	6.56	7.29	6.75	5.46	4.54	2122.5	5.8
LAS VEGAS	NV	LATITUDE: 36 DEGREES 5 MINUTES													
LATITUDE -15:		4.28	5.30	6.58	7.64	8.18	8.39	7.99	7.76	7.43	6.27	4.88	4.11	2398.8	6.6
LATITUDE :		4.86	5.72	6.74	7.37	7.52	7.56	7.35	7.49	7.64	6.85	5.59	4.79	2418.3	6.6
LATITUDE +15:		5.16	5.83	6.53	6.72	6.53	6.43	6.38	6.83	7.41	7.03	5.97	5.18	2311.4	6.3
LOVELOCK	NV	LATITUDE: 40 DEGREES 4 MINUTES													
LATITUDE -15:		3.86	4.88	6.20	7.26	7.95	8.32	8.67	8.40	7.76	6.41	4.64	3.70	2376.4	6.5
LATITUDE :		4.37	5.26	6.34	6.99	7.31	7.49	7.95	8.10	7.99	7.02	5.31	4.29	2387.9	6.5
LATITUDE +15:		4.62	5.34	6.13	6.38	6.35	6.38	6.89	7.38	7.75	7.21	5.67	4.63	2274.7	6.2
RENO	NV	LATITUDE: 39 DEGREES 30 MINUTES													
LATITUDE -15:		3.75	4.74	6.12	7.21	7.85	8.17	8.37	8.10	7.58	6.19	4.43	3.55	2316.2	6.3
LATITUDE :		4.24	5.10	6.26	6.95	7.22	7.37	7.68	7.82	7.80	6.78	5.06	4.12	2326.3	6.4
LATITUDE +15:		4.49	5.18	6.06	6.34	6.29	6.29	6.66	7.13	7.58	6.96	5.40	4.44	2215.6	6.1
TONOPAH	NV	LATITUDE: 38 DEGREES 4 MINUTES													
LATITUDE -15:		4.23	5.20	6.54	7.47	7.99	8.41	8.38	8.14	7.63	6.45	4.90	4.10	2419.3	6.6
LATITUDE :		4.81	5.62	6.71	7.20	7.35	7.58	7.69	7.86	7.85	7.06	5.62	4.78	2439.2	6.7
LATITUDE +15:		5.10	5.71	6.49	6.57	6.39	6.44	6.67	7.16	7.62	7.26	6.01	5.18	2331.1	6.4
WINNEMUCCA	NV	LATITUDE: 40 DEGREES 54 MINUTES													
LATITUDE -15:		3.26	4.27	5.48	6.59	7.37	7.79	8.36	7.97	7.35	5.81	4.01	3.18	2175.9	6.0
LATITUDE :		3.67	4.58	5.59	6.34	6.78	7.04	7.68	7.69	7.57	6.34	4.56	3.67	2178.2	6.0
LATITUDE +15:		3.87	4.63	5.40	5.79	5.92	6.02	6.67	7.02	7.35	6.50	4.85	3.95	2069.0	5.7
YUCCA FLATS	NV	LATITUDE: 36 DEGREES 57 MINUTES													
LATITUDE -15:		4.25	5.07	6.39	7.42	7.98	8.27	8.21	7.90	7.44	6.25	4.75	4.07	2375.0	6.5
LATITUDE :		4.84	5.47	6.55	7.17	7.35	7.47	7.55	7.63	7.66	6.84	5.45	4.75	2396.5	6.6
LATITUDE +15:		5.14	5.57	6.35	6.54	6.40	6.37	6.56	6.96	7.45	7.03	5.83	5.14	2292.6	6.3
ALBANY	NY	LATITUDE: 42 DEGREES 45 MINUTES													
LATITUDE -15:		2.08	2.77	3.57	4.40	4.86	5.24	5.35	4.98	4.32	3.41	2.09	1.68	1363.8	3.7
LATITUDE :		2.28	2.91	3.59	4.22	4.50	4.78	4.94	4.78	4.36	3.63	2.30	1.88	1344.6	3.7
LATITUDE +15:		2.37	2.90	3.43	3.84	3.96	4.15	4.34	4.35	4.18	3.65	2.39	1.97	1264.5	3.5
BINGHAMTON	NY	LATITUDE: 42 DEGREES 13 MINUTES													
LATITUDE -15:		1.84	2.20	3.05	4.07	4.63	5.09	5.14	4.71	4.13	3.20	1.80	1.48	1259.3	3.5
LATITUDE :		2.05	2.27	3.04	3.89	4.28	4.64	4.74	4.51	4.16	3.39	1.96	1.67	1236.8	3.4
LATITUDE +15:		2.15	2.23	2.89	3.55	3.77	4.03	4.17	4.11	3.98	3.40	2.01	1.77	1159.4	3.2
BUFFALO	NY	LATITUDE: 42 DEGREES 56 MINUTES													
LATITUDE -15:		1.65	2.27	3.18	4.34	4.96	5.47	5.52	5.04	4.25	3.26	1.78	1.43	1315.2	3.6
LATITUDE :		1.83	2.41	3.18	4.15	4.58	4.98	5.10	4.83	4.29	3.47	1.94	1.61	1291.3	3.5
LATITUDE +15:		1.92	2.42	3.03	3.78	4.03	4.32	4.48	4.40	4.12	3.48	2.00	1.71	1209.3	3.3
CENTRAL PARK	NY	LATITUDE: 40 DEGREES 47 MINUTES													
LATITUDE -15:		2.19	2.82	3.70	4.48	5.07	5.18	5.22	4.89	4.40	3.65	2.38	1.84	1395.3	3.8
LATITUDE :		2.40	2.96	3.72	4.29	4.69	4.73	4.83	4.69	4.44	3.88	2.63	2.05	1380.0	3.8
LATITUDE +15:		2.49	2.95	3.56	3.92	4.14	4.12	4.25	4.28	4.26	3.91	2.74	2.16	1302.0	3.6
LA GUARDIA	NY	LATITUDE: 40 DEGREES 46 MINUTES													
LATITUDE -15:		2.45	3.16	4.02	4.79	5.24	5.47	5.52	5.23	4.66	3.90	2.71	2.16	1501.8	4.1
LATITUDE :		2.71	3.34	4.05	4.60	4.85	4.98	5.10	5.02	4.72	4.17	3.02	2.44	1491.7	4.1
LATITUDE +15:		2.82	3.34	3.89	4.20	4.27	4.33	4.49	4.58	4.53	4.21	3.16	2.59	1412.3	3.9
MASSENA	NY	LATITUDE: 44 DEGREES 56 MINUTES													
LATITUDE -15:		1.84	2.56	3.63	4.48	5.03	5.39	5.45	4.99	4.23	3.17	1.83	1.43	1341.4	3.7
LATITUDE :		2.02	2.68	3.65	4.29	4.64	4.90	5.02	4.78	4.28	3.37	2.00	1.58	1316.2	3.6
LATITUDE +15:		2.09	2.67	3.49	3.91	4.08	4.25	4.41	4.35	4.10	3.39	2.07	1.66	1231.7	3.4
ROCHESTER	NY	LATITUDE: 43 DEGREES 7 MINUTES													
LATITUDE -15:		1.77	2.16	3.25	4.43	4.98	5.50	5.53	5.06	4.29	3.27	1.80	1.43	1325.1	3.6
LATITUDE :		1.97	2.23	3.25	4.24	4.60	5.00	5.09	4.85	4.33	3.47	1.96	1.61	1298.4	3.6
LATITUDE +15:		2.06	2.19	3.09	3.86	4.05	4.33	4.47	4.42	4.15	3.48	2.02	1.71	1213.6	3.3
SYRACUSE	NY	LATITUDE: 43 DEGREES 7 MINUTES													
LATITUDE -15:		1.68	2.21	3.19	4.38	4.89	5.38	5.46	5.01	4.32	3.26	1.77	1.45	1310.7	3.6
LATITUDE :		1.82	2.29	3.19	4.18	4.52	4.90	5.03	4.80	4.36	3.45	1.93	1.64	1283.0	3.5
LATITUDE +15:		1.86	2.25	3.03	3.81	3.97	4.24	4.42	4.37	4.18	3.46	1.98	1.74	1198.3	3.3

SITE ARRAY TILT		JAN	FEB	MAR	APR	MAY	JUN	JUL	AUG	SEP	OCT	NOV	DEC	ANNUAL TOTAL (KWH/SQ. M)	AVERAGE DAY (KWH/SQ. M)
AKRON-CANTON	OH	LATITUDE:		40 DEGREES		55 MINUTES									
LATITUDE -15:		1.79	2.49	3.41	4.45	5.17	5.58	5.54	5.28	4.64	3.71	2.23	1.54	1396.3	3.8
LATITUDE :		1.94	2.59	3.42	4.27	4.79	5.09	5.12	5.07	4.69	3.96	2.45	1.69	1372.9	3.8
LATITUDE +15:		1.99	2.57	3.27	3.90	4.22	4.42	4.50	4.63	4.50	3.99	2.55	1.76	1287.7	3.5
CINCINNATI	OH	LATITUDE:		39 DEGREES		4 MINUTES									
LATITUDE -15:		2.08	2.82	3.60	4.56	5.17	5.57	5.47	5.37	4.70	3.97	2.55	1.90	1454.5	4.0
LATITUDE :		2.26	2.95	3.62	4.37	4.78	5.08	5.06	5.15	4.76	4.23	2.82	2.11	1436.8	3.9
LATITUDE +15:		2.33	2.93	3.46	3.99	4.21	4.40	4.45	4.70	4.56	4.27	2.94	2.21	1353.6	3.7
CLEVELAND	OH	LATITUDE:		41 DEGREES		24 MINUTES									
LATITUDE -15:		1.80	2.29	3.26	4.43	5.21	5.59	5.67	5.24	4.54	3.56	2.05	1.55	1377.5	3.8
LATITUDE :		2.00	2.37	3.26	4.24	4.82	5.09	5.23	5.03	4.58	3.79	2.25	1.76	1353.7	3.7
LATITUDE +15:		2.10	2.33	3.11	3.87	4.24	4.41	4.59	4.58	4.40	3.82	2.32	1.87	1268.8	3.5
COLUMBUS	OH	LATITUDE:		40 DEGREES		0 MINUTES									
LATITUDE -15:		1.91	2.58	3.45	4.42	5.09	5.49	5.43	5.41	4.63	3.83	2.35	1.70	1410.4	3.9
LATITUDE :		2.07	2.68	3.46	4.23	4.71	5.00	5.01	5.19	4.68	4.08	2.59	1.87	1388.7	3.8
LATITUDE +15:		2.13	2.66	3.30	3.86	4.15	4.34	4.41	4.73	4.49	4.12	2.69	1.95	1304.3	3.6
DAYTON	OH	LATITUDE:		39 DEGREES		54 MINUTES									
LATITUDE -15:		2.07	2.79	3.62	4.60	5.26	5.68	5.60	5.43	4.77	3.93	2.48	1.80	1462.8	4.0
LATITUDE :		2.26	2.93	3.64	4.41	4.87	5.18	5.18	5.21	4.83	4.19	2.75	2.00	1444.8	4.0
LATITUDE +15:		2.33	2.92	3.48	4.03	4.30	4.49	4.56	4.76	4.64	4.23	2.86	2.10	1360.8	3.7
TOLEDO	OH	LATITUDE:		41 DEGREES		36 MINUTES									
LATITUDE -15:		1.87	2.67	3.57	4.55	5.33	5.69	5.74	5.36	4.69	3.79	2.25	1.60	1434.9	3.9
LATITUDE :		2.04	2.79	3.59	4.36	4.92	5.18	5.29	5.15	4.74	4.04	2.48	1.77	1411.6	3.9
LATITUDE +15:		2.10	2.77	3.43	3.98	4.33	4.49	4.65	4.69	4.55	4.07	2.58	1.85	1324.2	3.6
YOUNGSTOWN	OH	LATITUDE:		41 DEGREES		16 MINUTES									
LATITUDE -15:		1.77	2.21	3.13	4.18	4.91	5.33	5.37	4.97	4.35	3.48	1.99	1.53	1317.5	3.6
LATITUDE :		1.97	2.28	3.13	4.00	4.54	4.86	4.96	4.76	4.39	3.70	2.17	1.73	1294.8	3.5
LATITUDE +15:		2.06	2.25	2.98	3.65	4.00	4.21	4.36	4.34	4.20	3.72	2.24	1.84	1214.2	3.3
OKLAHOMA CITY	OK	LATITUDE:		35 DEGREES		24 MINUTES									
LATITUDE -15:		3.30	3.99	4.89	5.60	5.91	6.49	6.57	6.37	5.47	4.78	3.82	3.16	1837.9	5.0
LATITUDE :		3.69	4.25	4.96	5.39	5.47	5.91	6.07	6.13	5.57	5.15	4.31	3.61	1841.6	5.0
LATITUDE +15:		3.87	4.28	4.78	4.93	4.82	5.10	5.31	5.60	5.37	5.23	4.55	3.86	1755.8	4.8
TULSA	OK	LATITUDE:		36 DEGREES		12 MINUTES									
LATITUDE -15:		3.04	3.71	4.57	5.20	5.62	6.12	6.26	6.10	5.20	4.54	3.52	2.90	1729.3	4.7
LATITUDE :		3.38	3.94	4.62	5.00	5.21	5.58	5.78	5.87	5.28	4.87	3.95	3.30	1728.5	4.7
LATITUDE +15:		3.54	3.96	4.44	4.57	4.59	4.83	5.07	5.36	5.08	4.94	4.16	3.52	1645.0	4.5
ASTORIA	OR	LATITUDE:		46 DEGREES		9 MINUTES									
LATITUDE -15:		1.66	2.23	3.20	4.19	5.00	4.91	5.45	5.07	4.57	3.16	1.93	1.29	1300.5	3.6
LATITUDE :		1.85	2.32	3.20	3.99	4.61	4.47	5.01	4.85	4.63	3.35	2.13	1.42	1274.8	3.5
LATITUDE +15:		1.94	2.29	3.05	3.63	4.04	3.87	4.39	4.41	4.43	3.37	2.21	1.48	1191.7	3.3
BURNS	OR	LATITUDE:		43 DEGREES		35 MINUTES									
LATITUDE -15:		2.35	3.34	4.44	5.53	6.41	6.92	7.70	7.13	6.33	4.67	3.00	2.26	1831.1	5.0
LATITUDE :		2.61	3.54	4.49	5.31	5.91	6.27	7.08	6.86	6.48	5.04	3.38	2.57	1813.6	5.0
LATITUDE +15:		2.72	3.55	4.31	4.84	5.17	5.38	6.16	6.25	6.27	5.13	3.55	2.73	1707.5	4.7
MEDFORD	OR	LATITUDE:		42 DEGREES		22 MINUTES									
LATITUDE -15:		1.76	2.99	4.16	5.47	6.33	6.90	7.73	7.20	6.08	4.22	2.35	1.54	1728.7	4.7
LATITUDE :		1.91	3.15	4.19	5.25	5.83	6.25	7.10	6.93	6.22	4.53	2.60	1.70	1695.6	4.6
LATITUDE +15:		1.96	3.15	4.02	4.78	5.11	5.37	6.18	6.32	6.00	4.58	2.71	1.77	1581.9	4.3
NORTH BEND	OR	LATITUDE:		43 DEGREES		25 MINUTES									
LATITUDE -15:		2.03	2.89	3.90	5.03	5.79	6.04	6.57	6.03	5.24	3.85	2.57	1.91	1579.9	4.3
LATITUDE :		2.23	3.04	3.92	4.82	5.33	5.49	6.05	5.80	5.33	4.12	2.86	2.15	1557.5	4.3
LATITUDE +15:		2.31	3.04	3.75	4.39	4.68	4.74	5.28	5.28	5.12	4.16	3.00	2.27	1462.2	4.0
PENDLETON	OR	LATITUDE:		45 DEGREES		41 MINUTES									
LATITUDE -15:		1.62	2.57	3.95	5.07	6.02	6.52	7.54	6.91	6.00	4.18	2.23	1.49	1649.4	4.5
LATITUDE :		1.76	2.69	3.98	4.85	5.55	5.91	6.93	6.64	6.14	4.50	2.48	1.66	1618.1	4.4
LATITUDE +15:		1.81	2.68	3.81	4.41	4.86	5.08	6.03	6.05	5.93	4.56	2.59	1.74	1510.0	4.1
PORTLAND	OR	LATITUDE:		45 DEGREES		36 MINUTES									
LATITUDE -15:		1.59	2.25	3.31	4.37	5.18	5.37	6.38	5.70	4.68	3.16	1.88	1.46	1382.1	3.8
LATITUDE :		1.76	2.34	3.31	4.18	4.77	4.89	5.87	5.46	4.74	3.35	2.07	1.66	1353.9	3.7
LATITUDE +15:		1.84	2.31	3.16	3.80	4.19	4.23	5.13	4.97	4.55	3.37	2.15	1.77	1264.0	3.5
REDMOND	OR	LATITUDE:		44 DEGREES		16 MINUTES									
LATITUDE -15:		2.43	3.31	4.50	5.68	6.50	6.94	7.68	7.11	6.25	4.52	2.95	2.30	1833.4	5.0
LATITUDE :		2.70	3.50	4.55	5.45	5.98	6.28	7.05	6.84	6.39	4.88	3.31	2.62	1814.4	5.0
LATITUDE +15:		2.82	3.52	4.36	4.96	5.23	5.39	6.13	6.23	6.17	4.95	3.49	2.78	1706.8	4.7

The Solar Electric Independent Home Book

SITE ARRAY TILT	JAN	FEB	MAR	APR	MAY	JUN	JUL	AUG	SEP	OCT	NOV	DEC	ANNUAL TOTAL (KWH/SQ. M)	AVERAGE DAY (KWH/SQ. M)
SALEM OR LATITUDE: 44 DEGREES 55 MINUTES														
LATITUDE -15:	1.68	2.39	3.50	4.58	5.42	5.61	6.72	6.04	5.14	3.34	1.98	1.53	1461.0	4.0
LATITUDE :	1.88	2.49	3.51	4.38	5.00	5.10	6.18	5.81	5.22	3.55	2.18	1.74	1434.1	3.9
LATITUDE +15:	1.97	2.47	3.36	3.99	4.39	4.41	5.40	5.29	5.03	3.58	2.26	1.86	1340.9	3.7
ALLENTOWN PA LATITUDE: 40 DEGREES 39 MINUTES														
LATITUDE -15:	2.33	3.01	3.86	4.63	5.07	5.38	5.47	5.10	4.49	3.79	2.57	2.00	1452.4	4.0
LATITUDE :	2.57	3.17	3.89	4.44	4.69	4.91	5.05	4.89	4.54	4.04	2.85	2.24	1439.4	3.9
LATITUDE +15:	2.67	3.17	3.72	4.05	4.14	4.27	4.44	4.46	4.36	4.07	2.98	2.36	1360.1	3.7
ERIE PA LATITUDE: 42 DEGREES 5 MINUTES														
LATITUDE -15:	1.59	2.20	3.28	4.48	5.10	5.58	5.69	4.82	4.41	3.43	1.81	1.34	1332.2	3.6
LATITUDE :	1.75	2.27	3.28	4.28	4.71	5.08	5.24	4.61	4.45	3.64	1.96	1.50	1303.8	3.6
LATITUDE +15:	1.82	2.24	3.12	3.90	4.14	4.40	4.60	4.20	4.26	3.66	2.02	1.59	1217.3	3.3
HARRISBURG PA LATITUDE: 40 DEGREES 13 MINUTES														
LATITUDE -15:	2.34	3.02	3.86	4.62	5.11	5.47	5.45	5.10	4.59	3.80	2.60	2.07	1462.7	4.0
LATITUDE :	2.58	3.18	3.89	4.43	4.73	4.99	5.04	4.89	4.63	4.05	2.88	2.32	1449.2	4.0
LATITUDE +15:	2.68	3.18	3.72	4.04	4.17	4.33	4.43	4.46	4.44	4.08	3.01	2.45	1369.0	3.8
PHILADELPHIA PA LATITUDE: 39 DEGREES 53 MINUTES														
LATITUDE -15:	2.42	3.11	3.95	4.69	5.14	5.49	5.45	5.18	4.62	3.87	2.78	2.17	1488.2	4.1
LATITUDE :	2.67	3.28	3.98	4.50	4.76	5.01	5.04	4.98	4.67	4.13	3.10	2.44	1478.5	4.1
LATITUDE +15:	2.78	3.28	3.82	4.11	4.20	4.35	4.43	4.54	4.49	4.17	3.24	2.59	1400.4	3.8
PITTSBURGH PA LATITUDE: 40 DEGREES 30 MINUTES														
LATITUDE -15:	1.75	2.36	3.32	4.30	4.96	5.34	5.23	4.97	4.36	3.63	2.20	1.68	1343.8	3.7
LATITUDE :	1.88	2.45	3.33	4.12	4.59	4.87	4.83	4.77	4.41	3.87	2.42	1.91	1323.1	3.6
LATITUDE +15:	1.92	2.42	3.17	3.76	4.05	4.23	4.25	4.35	4.22	3.89	2.51	2.03	1243.0	3.4
WILKES-BARRE-SCRAN PA LATITUDE: 41 DEGREES 20 MINUTES														
LATITUDE -15:	1.98	2.69	3.54	4.39	4.92	5.33	5.41	5.00	4.37	3.70	2.18	1.67	1376.0	3.8
LATITUDE :	2.16	2.82	3.55	4.20	4.55	4.86	4.99	4.79	4.41	3.94	2.40	1.84	1355.7	3.7
LATITUDE +15:	2.22	2.80	3.39	3.83	4.01	4.22	4.39	4.37	4.22	3.97	2.49	1.93	1274.1	3.5
KOROR ISLAND PN LATITUDE: 7 DEGREES 20 MINUTES														
LATITUDE -15:	4.14	4.72	5.04	5.36	5.05	4.72	4.65	4.70	4.73	4.56	4.29	3.93	1698.9	4.7
LATITUDE :	4.60	5.03	5.15	5.23	4.77	4.41	4.40	4.59	4.82	4.87	4.78	4.45	1736.1	4.8
LATITUDE +15:	4.84	5.10	5.00	4.85	4.27	3.91	3.95	4.27	4.68	4.95	5.04	4.75	1690.1	4.6
KWAJALEIN ISLAND PN LATITUDE: 8 DEGREES 44 MINUTES														
LATITUDE -15:	4.69	5.32	5.57	5.50	5.23	5.20	5.16	5.34	5.01	4.64	4.35	4.28	1832.8	5.0
LATITUDE :	5.27	5.72	5.69	5.37	4.94	4.84	4.88	5.21	5.11	4.96	4.86	4.89	1876.2	5.1
LATITUDE +15:	5.58	5.82	5.53	4.97	4.41	4.27	4.36	4.82	4.96	5.04	5.13	5.26	1827.7	5.0
WAKE ISLAND PN LATITUDE: 19 DEGREES 17 MINUTES														
LATITUDE -15:	4.45	5.11	5.79	6.16	6.41	6.37	6.02	5.90	5.56	5.11	4.75	4.36	2007.1	5.5
LATITUDE :	5.01	5.48	5.91	5.97	5.98	5.84	5.61	5.71	5.66	5.49	5.38	5.02	2039.2	5.6
LATITUDE +15:	5.30	5.56	5.72	5.48	5.26	5.06	4.96	5.25	5.47	5.58	5.71	5.41	1969.7	5.4
SAN JUAN PR LATITUDE: 18 DEGREES 26 MINUTES														
LATITUDE -15:	4.33	4.96	5.69	5.95	5.68	5.67	5.86	5.80	5.33	4.89	4.47	4.05	1907.1	5.2
LATITUDE :	4.85	5.31	5.81	5.77	5.32	5.24	5.48	5.62	5.42	5.24	5.03	4.63	1938.3	5.3
LATITUDE +15:	5.12	5.38	5.63	5.31	4.72	4.58	4.85	5.17	5.24	5.32	5.33	4.97	1874.1	5.1
PROVIDENCE RI LATITUDE: 41 DEGREES 44 MINUTES														
LATITUDE -15:	2.30	2.95	3.72	4.53	5.14	5.38	5.25	4.95	4.43	3.77	2.48	2.01	1428.2	3.9
LATITUDE :	2.53	3.10	3.74	4.34	4.75	4.90	4.85	4.75	4.47	4.02	2.76	2.26	1415.1	3.9
LATITUDE +15:	2.64	3.10	3.58	3.96	4.18	4.26	4.27	4.33	4.29	4.06	2.88	2.39	1337.0	3.7
CHARLESTON SC LATITUDE: 32 DEGREES 54 MINUTES														
LATITUDE -15:	2.87	3.61	4.56	5.58	5.73	5.60	5.55	5.10	4.76	4.42	3.73	2.92	1656.9	4.5
LATITUDE :	3.17	3.82	4.62	5.38	5.33	5.14	5.15	4.91	4.82	4.73	4.19	3.31	1660.5	4.5
LATITUDE +15:	3.30	3.84	4.45	4.93	4.70	4.48	4.55	4.50	4.64	4.79	4.42	3.52	1585.6	4.3
COLUMBIA SC LATITUDE: 33 DEGREES 57 MINUTES														
LATITUDE -15:	3.01	3.76	4.66	5.64	5.85	5.91	5.68	5.50	4.97	4.57	3.76	3.01	1714.5	4.7
LATITUDE :	3.34	3.99	4.72	5.44	5.42	5.41	5.27	5.30	5.04	4.91	4.23	3.43	1719.3	4.7
LATITUDE +15:	3.49	4.02	4.55	4.98	4.78	4.69	4.65	4.85	4.85	4.98	4.47	3.65	1642.0	4.5
GREENVILLE-SPARTAN SC LATITUDE: 34 DEGREES 54 MINUTES														
LATITUDE -15:	2.93	3.66	4.60	5.50	5.67	5.82	5.65	5.50	4.88	4.50	3.66	2.84	1680.4	4.6
LATITUDE :	3.25	3.88	4.66	5.30	5.26	5.32	5.24	5.30	4.94	4.83	4.11	3.22	1683.3	4.6
LATITUDE +15:	3.39	3.90	4.48	4.85	4.64	4.62	4.62	4.85	4.75	4.90	4.34	3.43	1606.1	4.4
HURON SD LATITUDE: 44 DEGREES 23 MINUTES														
LATITUDE -15:	2.42	3.16	4.18	5.13	5.84	6.37	6.84	6.46	5.50	4.48	3.00	2.17	1692.0	4.6
LATITUDE :	2.70	3.34	4.22	4.91	5.38	5.78	6.29	6.20	5.60	4.83	3.37	2.46	1677.5	4.6
LATITUDE +15:	2.82	3.35	4.04	4.47	4.72	4.98	5.48	5.65	5.39	4.90	3.54	2.62	1582.2	4.3

SITE ARRAY TILT		JAN	FEB	MAR	APR	MAY	JUN	JUL	AUG	SEP	OCT	NOV	DEC	ANNUAL TOTAL (KWH/SQ. M)	AVERAGE DAY (KWH/SQ. M)
PIERRE	SD	LATITUDE:		44 DEGREES		23 MINUTES									
LATITUDE -15:		2.68	3.42	4.57	5.43	6.14	6.66	7.13	6.83	5.86	4.83	3.31	2.43	1806.6	4.9
LATITUDE :		3.00	3.63	4.63	5.21	5.65	6.03	6.56	6.57	5.98	5.22	3.74	2.78	1797.0	4.9
LATITUDE +15:		3.15	3.65	4.45	4.74	4.95	5.18	5.71	5.99	5.77	5.31	3.95	2.96	1699.3	4.7
RAPID CITY	SD	LATITUDE:		44 DEGREES		3 MINUTES									
LATITUDE -15:		2.73	3.55	4.65	5.33	5.88	6.46	6.95	6.70	5.92	4.85	3.43	2.63	1799.9	4.9
LATITUDE :		3.06	3.77	4.71	5.11	5.42	5.85	6.38	6.44	6.04	5.25	3.88	3.01	1794.2	4.9
LATITUDE +15:		3.20	3.79	4.52	4.65	4.75	5.03	5.56	5.86	5.83	5.34	4.10	3.21	1700.4	4.7
SIOUX FALLS	SD	LATITUDE:		43 DEGREES		34 MINUTES									
LATITUDE -15:		2.62	3.38	4.30	5.15	5.90	6.36	6.71	6.25	5.40	4.47	3.09	2.33	1704.4	4.7
LATITUDE :		2.92	3.59	4.34	4.94	5.44	5.77	6.18	6.00	5.49	4.82	3.48	2.65	1693.6	4.6
LATITUDE +15:		3.06	3.61	4.17	4.50	4.77	4.97	5.39	5.47	5.29	4.89	3.66	2.82	1601.3	4.4
CHATTANOOGA	TN	LATITUDE:		35 DEGREES		2 MINUTES									
LATITUDE -15:		2.48	3.16	4.04	5.01	5.33	5.56	5.34	5.27	4.63	4.21	3.16	2.41	1540.7	4.2
LATITUDE :		2.72	3.32	4.07	4.81	4.94	5.08	4.95	5.07	4.68	4.50	3.51	2.70	1533.0	4.2
LATITUDE +15:		2.81	3.31	3.90	4.40	4.36	4.42	4.37	4.63	4.49	4.54	3.68	2.85	1454.4	4.0
KNOXVILLE	TN	LATITUDE:		35 DEGREES		49 MINUTES									
LATITUDE -15:		2.48	3.21	4.12	5.19	5.56	5.77	5.56	5.41	4.84	4.31	3.15	2.40	1583.2	4.3
LATITUDE :		2.72	3.38	4.16	4.99	5.16	5.27	5.15	5.21	4.90	4.62	3.51	2.70	1576.1	4.3
LATITUDE +15:		2.82	3.38	4.00	4.57	4.55	4.58	4.54	4.76	4.71	4.67	3.68	2.86	1495.4	4.1
MEMPHIS	TN	LATITUDE:		35 DEGREES		3 MINUTES									
LATITUDE -15:		2.73	3.52	4.42	5.30	5.81	6.19	6.07	5.93	5.14	4.63	3.37	2.65	1697.8	4.7
LATITUDE :		3.01	3.71	4.46	5.10	5.38	5.64	5.62	5.71	5.21	4.97	3.76	2.99	1691.6	4.6
LATITUDE +15:		3.13	3.72	4.29	4.66	4.73	4.88	4.93	5.21	5.01	5.04	3.95	3.18	1605.3	4.4
NASHVILLE	TN	LATITUDE:		36 DEGREES		7 MINUTES									
LATITUDE -15:		2.30	3.06	3.91	5.01	5.62	5.94	5.83	5.66	4.91	4.32	2.95	2.18	1574.0	4.3
LATITUDE :		2.51	3.21	3.93	4.81	5.21	5.42	5.40	5.44	4.97	4.62	3.27	2.43	1559.6	4.3
LATITUDE +15:		2.59	3.20	3.77	4.40	4.59	4.69	4.74	4.97	4.77	4.67	3.42	2.55	1472.9	4.0
ABILENE	TX	LATITUDE:		32 DEGREES		26 MINUTES									
LATITUDE -15:		3.63	4.34	5.42	5.94	6.28	6.70	6.59	6.32	5.49	4.90	4.03	3.57	1923.9	5.3
LATITUDE :		4.06	4.63	5.51	5.72	5.82	6.10	6.09	6.09	5.58	5.27	4.54	4.10	1933.6	5.3
LATITUDE +15:		4.27	4.68	5.32	5.24	5.12	5.26	5.34	5.57	5.38	5.36	4.80	4.40	1848.2	5.1
AMARILLO	TX	LATITUDE:		35 DEGREES		14 MINUTES									
LATITUDE -15:		4.09	4.80	5.76	6.59	6.82	7.23	7.03	6.87	6.26	5.53	4.47	3.93	2112.4	5.8
LATITUDE :		4.62	5.16	5.88	6.35	6.30	6.56	6.48	6.63	6.40	5.99	5.09	4.56	2130.9	5.8
LATITUDE +15:		4.89	5.23	5.68	5.80	5.51	5.63	5.66	6.05	6.18	6.12	5.42	4.93	2041.5	5.6
AUSTIN	TX	LATITUDE:		30 DEGREES		18 MINUTES									
LATITUDE -15:		3.24	4.00	4.81	5.13	5.66	6.30	6.49	6.19	5.43	4.83	3.76	3.21	1797.6	4.9
LATITUDE :		3.58	4.25	4.87	4.95	5.26	5.75	6.01	5.98	5.52	5.18	4.20	3.64	1801.9	4.9
LATITUDE +15:		3.74	4.28	4.69	4.54	4.65	4.98	5.27	5.47	5.33	5.26	4.42	3.88	1719.6	4.7
BROWNSVILLE	TX	LATITUDE:		25 DEGREES		54 MINUTES									
LATITUDE -15:		3.20	3.86	4.78	5.52	5.97	6.48	6.84	6.44	5.58	4.96	3.74	3.08	1641.3	5.0
LATITUDE :		3.54	4.09	4.85	5.34	5.56	5.93	6.34	6.23	5.68	5.34	4.17	3.47	1843.4	5.1
LATITUDE +15:		3.69	4.12	4.68	4.91	4.91	5.14	5.56	5.71	5.49	5.43	4.38	3.68	1756.7	4.8
CORPUS CHRISTI	TX	LATITUDE:		27 DEGREES		46 MINUTES									
LATITUDE -15:		3.23	3.97	4.73	5.23	5.77	6.39	6.75	6.35	5.62	4.99	3.81	3.11	1825.4	5.0
LATITUDE :		3.57	4.21	4.80	5.05	5.37	5.84	6.25	6.14	5.72	5.37	4.26	3.51	1829.8	5.0
LATITUDE +15:		3.73	4.24	4.62	4.64	4.75	5.06	5.49	5.62	5.53	5.46	4.48	3.73	1746.0	4.8
DALLAS	TX	LATITUDE:		32 DEGREES		51 MINUTES									
LATITUDE -15:		3.21	3.91	4.86	5.23	5.82	6.47	6.54	6.30	5.47	4.77	3.73	3.19	1812.3	5.0
LATITUDE :		3.57	4.15	4.93	5.04	5.41	5.91	6.05	6.08	5.56	5.13	4.19	3.64	1816.3	5.0
LATITUDE +15:		3.74	4.18	4.75	4.62	4.77	5.11	5.31	5.56	5.37	5.21	4.42	3.89	1732.7	4.7
DEL RIO	TX	LATITUDE:		29 DEGREES		22 MINUTES									
LATITUDE -15:		3.57	4.27	5.30	5.43	5.64	6.16	6.34	6.20	5.32	4.87	3.99	3.47	1844.1	5.1
LATITUDE :		3.97	4.54	5.39	5.24	5.25	5.63	5.88	5.98	5.41	5.24	4.48	3.96	1855.9	5.1
LATITUDE +15:		4.17	4.58	5.21	4.80	4.64	4.88	5.17	5.48	5.22	5.32	4.73	4.23	1777.7	4.9
EL PASO	TX	LATITUDE:		31 DEGREES		48 MINUTES									
LATITUDE -15:		4.49	5.51	6.62	7.65	8.02	8.12	7.56	7.39	6.89	6.20	5.06	4.30	2367.7	6.5
LATITUDE :		5.10	5.96	6.79	7.39	7.39	7.34	6.97	7.14	7.07	6.77	5.80	5.01	2395.1	6.6
LATITUDE +15:		5.42	6.08	6.59	6.75	6.43	6.26	6.08	6.52	6.86	6.94	6.20	5.43	2297.8	6.3
FORT WORTH	TX	LATITUDE:		32 DEGREES		50 MINUTES									
LATITUDE -15:		3.13	3.90	4.82	5.20	5.83	6.53	6.65	6.41	5.59	4.84	3.74	3.13	1820.3	5.0
LATITUDE :		3.48	4.14	4.89	5.01	5.42	5.95	6.15	6.19	5.70	5.20	4.20	3.56	1823.3	5.0
LATITUDE +15:		3.64	4.17	4.71	4.59	4.78	5.15	5.39	5.66	5.50	5.29	4.43	3.80	1738.1	4.8

SITE ARRAY TILT		JAN	FEB	MAR	APR	MAY	JUN	JUL	AUG	SEP	OCT	NOV	DEC	ANNUAL TOTAL (KWH/SQ. M)	AVERAGE DAY (KWH/SQ. M)
HOUSTON	TX	LATITUDE:		29 DEGREES	59 MINUTES										
LATITUDE -15:		2.84	3.64	4.33	4.86	5.48	5.79	5.65	5.39	4.94	4.57	3.46	2.76	1635.1	4.5
LATITUDE :		3.11	3.85	4.38	4.69	5.11	5.31	5.25	5.21	5.01	4.90	3.85	3.10	1636.8	4.5
LATITUDE +15:		3.24	3.86	4.22	4.31	4.53	4.62	4.64	4.78	4.83	4.97	4.04	3.29	1561.7	4.3
KINGSVILLE	TX	LATITUDE:		27 DEGREES	31 MINUTES										
LATITUDE -15:		3.29	4.02	4.75	5.28	5.77	6.22	6.51	6.12	5.39	4.88	3.76	3.12	1800.0	4.9
LATITUDE :		3.63	4.26	4.81	5.10	5.37	5.69	6.04	5.92	5.48	5.24	4.20	3.53	1804.6	4.9
LATITUDE +15:		3.80	4.29	4.64	4.69	4.75	4.94	5.31	5.42	5.29	5.32	4.42	3.75	1722.6	4.7
LAREDO	TX	LATITUDE:		27 DEGREES	32 MINUTES										
LATITUDE -15:		3.46	4.15	5.02	5.49	6.03	6.33	6.58	6.41	5.67	4.94	3.79	3.28	1862.0	5.1
LATITUDE :		3.84	4.40	5.10	5.30	5.61	5.79	6.10	6.19	5.78	5.31	4.23	3.72	1868.4	5.1
LATITUDE +15:		4.02	4.44	4.92	4.87	4.95	5.02	5.36	5.67	5.58	5.40	4.45	3.96	1784.3	4.9
LUBBOCK	TX	LATITUDE:		33 DEGREES	39 MINUTES										
LATITUDE -15:		4.25	5.04	6.18	7.04	7.40	7.69	7.44	7.19	6.39	5.65	4.68	4.05	2222.0	6.1
LATITUDE :		4.82	5.43	6.32	6.80	6.83	6.97	6.86	6.94	6.54	6.13	5.34	4.71	2242.0	6.1
LATITUDE +15:		5.11	5.52	6.11	6.21	5.96	5.97	5.98	6.34	6.33	6.27	5.70	5.09	2147.3	5.9
LUFKIN	TX	LATITUDE:		31 DEGREES	14 MINUTES										
LATITUDE -15:		2.99	3.84	4.65	5.20	5.76	6.24	6.18	5.98	5.20	4.95	3.73	3.02	1758.3	4.8
LATITUDE :		3.30	4.06	4.71	5.01	5.35	5.70	5.73	5.77	5.28	5.33	4.17	3.42	1760.2	4.8
LATITUDE +15:		3.44	4.09	4.53	4.59	4.72	4.93	5.04	5.28	5.08	5.41	4.39	3.64	1678.1	4.6
MIDLAND-ODESSA	TX	LATITUDE:		31 DEGREES	56 MINUTES										
LATITUDE -15:		4.30	5.12	6.36	7.08	7.50	7.76	7.36	7.15	6.37	5.71	4.75	4.17	2241.4	6.1
LATITUDE :		4.87	5.51	6.52	6.84	6.93	7.04	6.80	6.91	6.52	6.21	5.42	4.84	2264.1	6.2
LATITUDE +15:		5.16	5.61	6.32	6.26	6.05	6.02	5.94	6.31	6.31	6.36	5.78	5.24	2170.7	5.9
PORT ARTHUR	TX	LATITUDE:		29 DEGREES	57 MINUTES										
LATITUDE -15:		2.95	3.78	4.53	5.14	5.78	6.13	5.70	5.56	5.14	4.75	3.58	2.87	1701.5	4.7
LATITUDE :		3.25	4.00	4.58	4.96	5.38	5.61	5.30	5.37	5.21	5.11	4.00	3.23	1704.2	4.7
LATITUDE +15:		3.38	4.02	4.42	4.56	4.75	4.87	4.69	4.92	5.03	5.19	4.20	3.43	1626.2	4.5
SAN ANGELO	TX	LATITUDE:		31 DEGREES	22 MINUTES										
LATITUDE -15:		3.72	4.38	5.48	5.94	6.27	6.63	6.54	6.32	5.48	4.92	4.10	3.61	1929.2	5.3
LATITUDE :		4.16	4.67	5.58	5.73	5.81	6.04	6.05	6.10	5.57	5.29	4.62	4.14	1940.3	5.3
LATITUDE +15:		4.37	4.72	5.38	5.25	5.11	5.21	5.31	5.58	5.37	5.38	4.88	4.45	1856.2	5.1
SAN ANTONIO	TX	LATITUDE:		29 DEGREES	32 MINUTES										
LATITUDE -15:		3.32	4.08	4.85	5.15	5.85	6.30	6.54	6.23	5.52	4.85	3.79	3.24	1818.1	5.0
LATITUDE :		3.68	4.33	4.92	4.96	5.44	5.75	6.06	6.01	5.61	5.21	4.24	3.68	1823.7	5.0
LATITUDE +15:		3.85	4.37	4.74	4.56	4.80	4.98	5.32	5.51	5.42	5.29	4.47	3.92	1741.4	4.8
SHERMAN	TX	LATITUDE:		33 DEGREES	43 MINUTES										
LATITUDE -15:		3.15	3.82	4.69	5.19	5.71	6.42	6.41	6.26	5.48	4.80	3.73	3.09	1789.0	4.9
LATITUDE :		3.50	4.05	4.75	5.00	5.30	5.85	5.93	6.03	5.58	5.16	4.19	3.53	1792.4	4.9
LATITUDE +15:		3.66	4.08	4.57	4.58	4.68	5.06	5.20	5.52	5.38	5.25	4.43	3.76	1709.5	4.7
WACO	TX	LATITUDE:		31 DEGREES	37 MINUTES										
LATITUDE -15:		3.18	3.95	4.84	5.17	5.48	6.41	6.56	6.31	5.47	4.78	3.72	3.20	1799.0	4.9
LATITUDE :		3.53	4.19	4.91	4.98	5.09	5.85	6.08	6.09	5.56	5.14	4.17	3.64	1803.3	4.9
LATITUDE +15:		3.69	4.22	4.73	4.57	4.50	5.06	5.33	5.57	5.37	5.22	4.39	3.89	1720.8	4.7
WICHITA FALLS	TX	LATITUDE:		33 DEGREES	58 MINUTES										
LATITUDE -15:		3.47	4.18	5.10	5.70	6.22	6.73	6.68	6.39	5.57	4.91	3.93	3.39	1895.9	5.2
LATITUDE :		3.87	4.46	5.18	5.49	5.77	6.13	6.17	6.16	5.67	5.29	4.44	3.90	1903.8	5.2
LATITUDE +15:		4.07	4.50	5.00	5.03	5.07	5.29	5.41	5.63	5.48	5.39	4.70	4.18	1818.4	5.0
BRYCE CANYON	UT	LATITUDE:		37 DEGREES	42 MINUTES										
LATITUDE -15:		4.14	4.97	6.13	7.05	7.60	8.03	7.51	7.14	7.09	6.10	4.74	3.98	2267.0	6.2
LATITUDE :		4.70	5.36	6.27	6.80	7.01	7.26	6.92	6.89	7.28	6.67	5.43	4.63	2288.9	6.3
LATITUDE +15:		4.99	5.45	6.07	6.21	6.11	6.20	6.03	6.29	7.06	6.84	5.80	5.02	2192.4	6.0
CEDAR CITY	UT	LATITUDE:		37 DEGREES	42 MINUTES										
LATITUDE -15:		3.97	4.71	5.94	6.92	7.65	8.19	7.76	7.43	7.29	6.07	4.61	3.78	2262.1	6.2
LATITUDE :		4.50	5.06	6.07	6.67	7.05	7.39	7.14	7.17	7.50	6.63	5.27	4.38	2278.0	6.2
LATITUDE +15:		4.77	5.14	5.87	6.09	6.14	6.30	6.22	6.54	7.28	6.80	5.63	4.74	2176.3	6.0
SALT LAKE CITY	UT	LATITUDE:		40 DEGREES	46 MINUTES										
LATITUDE -15:		2.96	4.07	5.40	6.32	7.36	7.76	8.08	7.61	7.05	5.64	3.86	2.84	2100.2	5.8
LATITUDE :		3.31	4.35	5.50	6.08	6.78	7.02	7.42	7.34	7.25	6.14	4.38	3.26	2096.1	5.7
LATITUDE +15:		3.48	4.39	5.31	5.55	5.92	6.00	6.45	6.69	7.03	6.29	4.65	3.48	1986.1	5.4
NORFOLK	VA	LATITUDE:		36 DEGREES	54 MINUTES										
LATITUDE -15:		2.83	3.55	4.50	5.46	5.83	6.07	5.72	5.48	4.93	4.22	3.50	2.77	1669.8	4.6
LATITUDE :		3.14	3.76	4.55	5.26	5.40	5.53	5.30	5.27	5.00	4.52	3.94	3.15	1668.1	4.6
LATITUDE +15:		3.28	3.78	4.38	4.81	4.76	4.79	4.66	4.82	4.81	4.58	4.15	3.36	1587.1	4.3

SITE ARRAY TILT		JAN	FEB	MAR	APR	MAY	JUN	JUL	AUG	SEP	OCT	NOV	DEC	ANNUAL TOTAL (KWH/SQ. M)	AVERAGE DAY (KWH/SQ. M)
RICHMOND	VA	LATITUDE:	37 DEGREES	30 MINUTES											
LATITUDE -15:		2.65	3.35	4.25	5.10	5.44	5.68	5.47	5.22	4.77	4.05	3.16	2.52	1572.6	4.3
LATITUDE :		2.93	3.54	4.29	4.90	5.04	5.18	5.07	5.01	4.83	4.32	3.53	2.85	1567.2	4.3
LATITUDE +15:		3.05	3.54	4.12	4.48	4.44	4.50	4.46	4.58	4.64	4.36	3.71	3.03	1488.6	4.1
ROANOKE	VA	LATITUDE:	37 DEGREES	19 MINUTES											
LATITUDE -15:		2.78	3.44	4.35	5.15	5.45	5.71	5.54	5.28	4.81	4.25	3.31	2.64	1604.6	4.4
LATITUDE :		3.08	3.64	4.39	4.95	5.05	5.21	5.13	5.08	4.86	4.55	3.71	2.99	1601.9	4.4
LATITUDE +15:		3.22	3.64	4.22	4.52	4.45	4.52	4.51	4.64	4.67	4.60	3.90	3.18	1523.8	4.2
BURLINGTON	VT	LATITUDE:	44 DEGREES	28 MINUTES											
LATITUDE -15:		1.78	2.46	3.45	4.30	4.89	5.24	5.36	4.94	4.21	3.17	1.71	1.54	1311.3	3.6
LATITUDE :		1.94	2.57	3.46	4.11	4.51	4.77	4.94	4.73	4.24	3.36	1.86	1.76	1286.8	3.5
LATITUDE +15:		2.00	2.55	3.30	3.74	3.97	4.13	4.33	4.30	4.06	3.37	1.92	1.87	1204.5	3.3
OLYMPIA	WA	LATITUDE:	46 DEGREES	58 MINUTES											
LATITUDE -15:		1.39	2.04	3.14	4.21	5.10	5.14	6.00	5.28	4.51	2.80	1.67	1.27	1297.0	3.6
LATITUDE :		1.54	2.11	3.14	4.02	4.70	4.67	5.52	5.06	4.56	2.96	1.83	1.44	1266.6	3.5
LATITUDE +15:		1.60	2.08	2.99	3.65	4.12	4.05	4.83	4.60	4.37	2.97	1.89	1.53	1179.3	3.2
SEATTLE-TACOMA	WA	LATITUDE:	47 DEGREES	27 MINUTES											
LATITUDE -15:		1.38	2.02	3.19	4.35	5.36	5.46	7.10	5.56	4.50	2.95	1.69	1.23	1366.3	3.7
LATITUDE :		1.52	2.10	3.19	4.16	4.93	4.96	6.51	5.32	4.55	3.13	1.86	1.39	1330.4	3.6
LATITUDE +15:		1.58	2.07	3.04	3.78	4.32	4.28	5.67	4.84	4.36	3.14	1.92	1.48	1234.3	3.4
SPOKANE	WA	LATITUDE:	47 DEGREES	38 MINUTES											
LATITUDE -15:		1.54	2.65	4.05	5.10	6.03	6.34	7.47	6.81	5.88	4.03	2.15	1.38	1628.7	4.5
LATITUDE :		1.68	2.80	4.09	4.88	5.55	5.74	6.86	6.55	6.01	4.33	2.39	1.54	1596.9	4.4
LATITUDE +15:		1.73	2.79	3.92	4.44	4.85	4.94	5.96	5.96	5.80	4.39	2.50	1.61	1489.2	4.1
WHIDBEY ISLAND	WA	LATITUDE:	48 DEGREES	21 MINUTES											
LATITUDE -15:		1.60	2.29	3.53	4.57	5.52	5.51	6.24	5.50	4.69	3.02	1.92	1.27	1392.0	3.8
LATITUDE :		1.79	2.40	3.55	4.36	5.07	5.00	5.73	5.27	4.75	3.21	2.13	1.41	1361.1	3.7
LATITUDE +15:		1.88	2.38	3.39	3.96	4.44	4.32	5.00	4.79	4.56	3.22	2.22	1.48	1267.9	3.5
GREEN BAY	WI	LATITUDE:	44 DEGREES	29 MINUTES											
LATITUDE -15:		2.20	3.06	4.14	4.81	5.35	5.78	5.89	5.47	4.62	3.57	2.30	1.79	1491.8	4.1
LATITUDE :		2.43	3.24	4.18	4.60	4.94	5.25	5.42	5.25	4.67	3.81	2.55	2.02	1472.1	4.0
LATITUDE +15:		2.53	3.24	4.01	4.19	4.33	4.54	4.75	4.78	4.48	3.84	2.66	2.13	1384.0	3.8
LA CROSSE	WI	LATITUDE:	43 DEGREES	52 MINUTES											
LATITUDE -15:		2.32	3.22	4.10	4.75	5.33	5.78	5.93	5.62	4.68	3.73	2.42	1.87	1514.7	4.1
LATITUDE :		2.58	3.41	4.14	4.55	4.92	5.25	5.46	5.39	4.74	3.99	2.69	2.10	1498.5	4.1
LATITUDE +15:		2.69	3.42	3.96	4.15	4.33	4.54	4.79	4.91	4.55	4.02	2.81	2.22	1412.3	3.9
MADISON	WI	LATITUDE:	43 DEGREES	8 MINUTES											
LATITUDE -15:		2.48	3.37	4.21	4.64	5.41	5.90	6.01	5.73	4.89	3.92	2.42	1.94	1550.6	4.2
LATITUDE :		2.75	3.57	4.25	4.44	4.99	5.36	5.54	5.50	4.95	4.20	2.68	2.18	1534.5	4.2
LATITUDE +15:		2.87	3.58	4.07	4.04	4.38	4.63	4.85	5.01	4.75	4.24	2.80	2.30	1446.4	4.0
MILWAUKEE	WI	LATITUDE:	42 DEGREES	57 MINUTES											
LATITUDE -15:		2.24	3.01	4.01	4.79	5.50	5.99	6.11	5.77	4.92	3.88	2.52	1.85	1540.7	4.2
LATITUDE :		2.47	3.18	4.04	4.59	5.08	5.45	5.63	5.55	4.99	4.16	2.80	2.08	1522.9	4.2
LATITUDE +15:		2.57	3.18	3.87	4.18	4.46	4.71	4.94	5.06	4.80	4.20	2.93	2.19	1433.8	3.9
CHARLESTON	WV	LATITUDE:	38 DEGREES	22 MINUTES											
LATITUDE -15:		2.02	2.64	3.52	4.41	5.06	5.39	5.20	4.95	4.52	3.84	2.63	1.89	1402.6	3.8
LATITUDE :		2.20	2.75	3.53	4.23	4.69	4.92	4.81	4.75	4.56	4.09	2.91	2.10	1386.2	3.8
LATITUDE +15:		2.26	2.73	3.37	3.86	4.14	4.28	4.24	4.34	4.38	4.12	3.04	2.19	1307.2	3.6
HUNTINGTON	WV	LATITUDE:	38 DEGREES	22 MINUTES											
LATITUDE -15:		2.16	2.87	3.74	4.72	5.28	5.59	5.46	5.17	4.65	3.98	2.75	2.04	1474.5	4.0
LATITUDE :		2.36	3.00	3.76	4.53	4.89	5.10	5.05	4.96	4.70	4.24	3.05	2.28	1459.7	4.0
LATITUDE +15:		2.44	2.99	3.60	4.14	4.31	4.43	4.45	4.53	4.51	4.28	3.19	2.40	1378.1	3.8
CASPER	WY	LATITUDE:	42 DEGREES	55 MINUTES											
LATITUDE -15:		3.49	4.41	5.49	6.23	6.88	7.59	7.95	7.62	6.85	5.55	4.05	3.31	2114.3	5.8
LATITUDE :		3.95	4.74	5.60	5.99	6.34	6.86	7.30	7.36	7.04	6.04	4.63	3.84	2121.9	5.8
LATITUDE +15:		4.18	4.80	5.40	5.46	5.54	5.88	6.35	6.71	6.83	6.19	4.93	4.15	2021.2	5.5
ROCK SPRINGS	WY	LATITUDE:	41 DEGREES	36 MINUTES											
LATITUDE -15:		3.63	4.65	5.78	6.53	7.31	7.80	7.96	7.60	7.10	5.83	4.24	3.50	2190.3	6.0
LATITUDE :		4.11	5.01	5.90	6.28	6.73	7.05	7.31	7.33	7.30	6.36	4.84	4.07	2200.7	6.0
LATITUDE +15:		4.35	5.08	5.70	5.73	5.87	6.03	6.35	6.69	7.08	6.52	5.16	4.39	2098.0	5.7
SHERIDAN	WY	LATITUDE:	44 DEGREES	46 MINUTES											
LATITUDE -15:		2.65	3.41	4.58	5.17	5.88	6.55	7.31	6.90	5.91	4.61	3.14	2.46	1784.6	4.9
LATITUDE :		2.96	3.62	4.64	4.95	5.42	5.94	6.72	6.64	6.04	4.98	3.54	2.82	1775.2	4.9
LATITUDE +15:		3.11	3.64	4.46	4.51	4.75	5.11	5.85	6.05	5.83	5.07	3.74	3.01	1678.9	4.6

SEASONAL NOONTIME SUN ANGLE
and
SEASONAL ARRAY ANGLE ADJUSTMENT

(ANGLES MEASURED IN DEGREES)

LATITUDE (IN DEGREES)	48	44	40	36	32	32
SUMMER SEASON (May, June, July)						
MAXIMUM SUN ANGLE ABOVE HORIZON	65.5	69.5	73.5	77.5	81.5	85.5
RECOMMENDED ARRAY ANGLE ABOVE HORIZON, LATITUDE – 15 DEGREES	33	29	25	21	17	13
WINTER SEASON (Nov., Dec., Jan.)						
MINIMUM SUN ANGLE ABOVE HORIZON	18.5	22.5	26.5	30.5	34.5	38.5
RECOMMENDED ARRAY ANGLE ABOVE HORIZON, LATITUDE + 15 DEGREES	63	59	55	51	47	43
SPRING & FALL SEASONS (Feb., March, April) (Aug., Sept., Oct.)						
AVERAGE SUN ANGLE ABOVE HORIZON	42	46	50	54	58	62
RECOMMENDED ARRAY ANGLE ABOVE HORIZON, EQUAL TO LATITUDE	48	44	40	36	32	28

INDEX

Index